The Chinese Alligator

The CHINESE

ALLIGATOR

Ecology, Behavior, Conservation, and Culture

John Thorbjarnarson
and Xiaoming Wang

Foreword by
George B. Schaller

THE JOHNS HOPKINS UNIVERITY PRESS

Baltimore

The Johns Hopkins University Press
2715 North Charles Street
Baltimore, Maryland 21218-4363
www.press.jhu.edu

Library of Congress Cataloging-in-Publication Data

Thorbjarnarson, John B.
The Chinese alligator : ecology, behavior, conservation, and culture /
John Thorbjarnarson and Xiaoming Wang ; foreword by George B. Schaller.
p.   cm.
Includes bibliographical references and index.
ISBN-13: 978-0-8018-9348-3 (hardcover : alk. paper)
ISBN-10: 0-8018-9348-8 (hardcover : alk. paper)
1. Chinese alligator. I. Wang, Xiaoming, 1963– II. Schaller, George B. III. Title.
QL666.C925T485 2009
597.98'4—dc22   2009020313

A catalog record for this book is available from the British Library.

*Special discounts are available for bulk purchases of this book. For more information,
please contact Special Sales at 410-516-6936 or specialsales@press.jhu.edu.*

The Johns Hopkins University Press uses environmentally friendly book materials,
including recycled text paper that is composed of at least 30 percent post-consumer waste,
whenever possible. All of our book papers are acid-free, and our jackets and covers are
printed on paper with recycled content.

To my parents, for all their love and support
JT

To my wife Yi Li and my daughter Ouya
XW

Contents

Foreword

With the dramatic increase in the world's human population during past decades, more and more natural habitat has been destroyed or degraded. About a third of the earth's land surface is already devoted to agriculture. As a result, many plant and animal species now survive only in fragments of habitat surrounded by a human-dominated environment. Conservationists have traditionally attempted to help save a country's natural heritage by promoting the creation of protected areas. However, most such areas lack the space and variety of habitats to maintain populations of all species, especially the larger ones, on a long-term basis. One proposed solution is to create corridors that connect habitat fragments. Another is to manage a whole landscape with an optimal mix of conservation and regulated development ranging from protected core areas to those designated for human use. It is clear that in the near future many species will depend for their survival on well-planned and well-executed conservation strategies that also take into account the aspirations, values, and livelihoods of local people. Nature and human needs must be linked. Such is the current model of the conservation community.

When John Thorbjarnarson and Xiaoming Wang began their study of the threatened Chinese alligator in the late 1990s, they faced a new and disconcerting reality. Instead of an increasing population, as they had been told, they found the *tu long*, the earth dragon, as the Chinese refer to the alligator, almost as rare as the mythical dragon. Hunting and habitat destruction had made the alligator functionally extinct in the wild, the animals reduced to a few scattered individuals in rice fields, drainage canals, and polluted village ponds. In the past hundred years, half of the world's wetlands have been lost, but here, in the fertile floodplains of the Yangtze River, all habitats have been converted to agriculture.

To restore a species to the wild after its habitat has vanished is an interesting challenge. *The Chinese Alligator* describes the effort to study the last alligators, noting their secretive nature and ability to adapt to marginal habitats. The American alligator breeds readily in captivity and easily returns to a free-living life. China had established breeding centers with eggs collected from some of the last nests in the wild, and there were now over ten thousand alligators in captivity. Reintroduction and restoration of the species to the wild was an obvious goal, the authors reasoned. They searched widely for potential wetland sites. They discussed with local communities the possibility of creating a narrow buffer of natural vegetation around certain ponds to improve nesting success. These and other ideas depended, of course, on obtaining captive alligators for restoration from government breeding centers. Here implementation stumbled over cultural perceptions.

The Chinese alligator was designated one of fifteen key species by the government, all of them rare, such as the giant panda and Przewalski's gazelle, and in need of special protection. Captive propagation is a widely used and legitimate method of protecting a species until it can be returned to the wild. But, as Richard Harris wrote in his book *Wildlife Conservation in China*, "captive breeding is assumed to be an end in itself" in China. Protecting the animal from the rigors of nature seems to be the idea. I attended an international workshop on the reintroduction of giant pandas a decade ago, yet no program is so far evident. As an initial step in alligator reintroduction, the authors managed the first small trial releases in spite of the government's lack of interest. There is an old Chinese saying, one very apt here: "When the wind changes direction, there are those who build walls and those who build windmills." Instead of walls for animals, China needs an innovative and flexible system of wildlife and habitat management, not just for the alligator but for all species.

Despite such unique challenges, the authors have retained their optimism, persistence, and commitment to reintroducing the Chinese alligator to its traditional and natural home. In a splendid example of international cooperation, they are at work on a grand master plan with the goal of having viable populations in restored wetlands within thirty years. They know that there is room for both alligators and people. And, as they point out, the earth dragon serves as a symbol of China's wetlands and China's ancient culture.

The Chinese Alligator shows with detail, insight, and vision that an attempt must be made to save species, no matter how complex the problems. Too many plants and animals spiral toward extinction quietly and without someone to mourn their fate. This book serves as an example, a guide, showing that with meticulous science, tenacity, cultural awareness, and hope, any species can be offered a future. The compelling message is that we must conserve the diversity of life and maintain ecological processes on behalf of all living beings, including the human species.

George B. Schaller
Senior Conservation Scientist, Wildlife Conservation Society
Vice President, Panthera

Preface

The Chinese have an idea that the alligator is wonderfully tenacious of life.

—A. A. Fauvel, "Alligators in China," 1879

In July and August 1999 we collaborated with the Anhui Forestry Bureau to conduct a systematic survey of the entire population of wild Chinese alligators. Our survey started in a confident mood, with a list of areas to visit and a healthy degree of optimism that the five-county region that constitutes the National Chinese Alligator Reserve would harbor at least a few suitable areas for the long-term conservation of this remarkable but little known species. Neither of us was prepared for the type of survey we were to undertake. One of us (JT) had studied alligators and crocodiles around the world, but never in an area so thoroughly converted to agriculture; a landscape of endless rice fields and drainage canals. The other (XW) had worked for many years in his native China, but primarily on the forested slopes of western Sichuan Province and on the windswept barrens of the Tibetan Plateau. When looking for Chinese alligators, we found ourselves in a densely populated area a day's journey to the west of Shanghai, the largest city in the world's most populous nation, spending most of our time driving from site to site, meeting with local officials, and talking with farmers.

Not too long ago these fertile eastern plains of China were home to a remarkable diversity of habitats and an impressive collection of wildlife, ranging from elephants and rhinos to the world's largest salamander. However, the lower Yangtze River valley was also a crucible for one of the world's greatest civilizations and over the course of the last two millennia had been transformed into China's breadbasket, with predictable consequences for its biodiversity. Particularly hard hit were the species that lived in the low, fertile plains and in the margins of rivers and lakes, the areas most heavily impacted by people. In 2006 the sad news emerged that one of the last animal holdouts, the *baiji*, or Yangtze white-flag dolphin (*Lipotes vexillifer*), had likely become extinct (Turvey et al. 2007). Today, apart from birds, which have the advantage of a highly mobile nature, little remains of the native wildlife that once inhabited the area's forests and wetlands. Of the larger mammals only the Chinese water deer (*Hydropotes inermis*) and the Eurasian otter (*Lutra lutra*) survive, in very small numbers. Another survivor, perhaps surprisingly, is the Chinese alligator (*Alligator sinensis*).

Figure P.1. A large adult Chinese alligator basking on a warm spring morning.

The survival of the Chinese alligator in the face of widespread persecution and the almost total loss of its natural habitat is seen by some as indicative of its tenacious hold on life. Actually, it is a reflection of the species' ecological and behavioral plasticity, its ability to adapt to and survive in marginal habitats, and its secretive nature. Pushed before a wave of rice cultivation that is more than 7,000 years in the making, they have retreated to their last sanctuary, a handful of small ponds in southern Anhui where the Yangtze floodplain runs up into the Huangshan Mountains. In a very real sense, alligators have become refugees in the agricultural landscape of the lower Yangtze River valley.

The Chinese alligator is one of the real-life inspirations for the legend of the Chinese dragon (Chapter 4). In southern Anhui Province, where the last alligators remain, local farmers still refer to them as *tu long*, or earth dragon. Its mythical counterpart, the legendary Chinese dragon, or *long*, is firmly embedded in the imagination and symbolism of the Chinese nation. China is experiencing a period of unprecedented growth and will emerge as a nation transformed. Its identity as the dragon nation will ensure that this powerful symbol will flourish. However, the survival of the Chinese alligator outside of breeding centers is less than certain. By virtue of its tiny and highly fragmented wild population, the Chinese alligator could become the first crocodilian to become extinct in the wild within historical times.

Nevertheless, we firmly believe that the Chinese alligator can be saved, even in the intimidating environment of today's Eastern China. Like other crocodilians, it has great potential as a flagship species, an ambassador for efforts to protect and restore wetlands in the lower and middle Yangtze River basin. Some of the same biological characteristics that have contributed to the alligator's survival will also help speed population recovery and restoration. Like many reptiles, alligators are relatively easy to breed in captivity, but even when they are raised in the relative comfort of a zoo they readily adapt to natural habitats when released (unlike most birds and mammals that generally require extensive periods of training or adaptation). Alligators are extremely adaptable and can live in even highly disturbed areas. Their land require-ments are modest, a consequence of having a low metabolic rate, so they do not need large wildlife reserves to establish viable populations. Chinese alligators are small (for crocodilians) and extremely secretive. They dig complex, underground burrows that serve as refuges from people and from environmental extremes. Their reptilian physiology allows them to remain hidden and inactive for long periods of time, and their food demands are not high.

We have written this book at what we hope is the turning of the tide. The alliga-tor has been losing ground to the burgeoning Chinese nation for at least 7,000 years. Over the last two millennia in particular, the agricultural frontier has swept through the entire historic distribution of the Chinese alligator. The last fragmented remains of a formerly robust wild population are now huddled in small ponds, literally in the shadows of the farmhouses that dot the former floodplains. In captivity, however, the Chinese alligator has thrived, and this source provides repopulation options for its future. The Chinese government has begun restoring animals to the wild using farm-bred alligators. Trial releases of captive-born alligators into small pockets of suitable habitat have shown that they quickly adjust to their new environment. Thus this technique can be used as the basis for reestablishing self-supporting populations outside of the breeding centers. The biggest challenge will be rehabilitating the alliga-tors' habitat. The survival of this species in the wild will rest upon the success of wet-lands conservation programs in China, the importance of which has been recently, and dramatically, highlighted by devastating floods along the Yangtze and other riv-ers in China. We are under no illusions that Chinese alligators will ever again be a widespread resident of the lower Yangtze. That era has passed. However, a well-planned program of wetlands conservation that is coordinated with wildlife restora-tion programs can return the alligator, and other native fauna, to representative sites in Eastern China. Despite its present dilemma, the Chinese alligator is a species with a potentially promising, perhaps even dramatically successful, future. It is with this hope that we have written this book.

Acknowledgments

We owe a tremendous debt of gratitude to a large number of people without whom we could not have carried out our work with Chinese alligators or dared to write this book. The list of our Chinese friends and collaborators is long. Our work was authorized by the State Forestry Administration (SFA) in Beijing, and for this we are grateful to Wei Wang, deputy director of the Department of Wildlife Conservation and Nature Reserve Management; Weisheng Wang, director of the Wild Animal Management Division at SFA; and Xiangdong Ruan, deputy director of the Wild Animal Management Division. At the national level, conservation activities for Chinese alligators are coordinated by Hongxing Jiang at the Research Institute of Forestry Ecology, Environment and Protection in the Chinese Academy of Forestry, and we appreciate his continued support. In Anhui Province, we worked with a large number of people in the Forestry Bureau, facilitated through the years by Kuichu Shi, formerly the director of the Division of Wildlife Conservation. In recent years, that position has been filled by our good friend Changming Gu, director of the Wildlife Conservation Station in Anhui Province, whose enthusiasm for alligators is always evident. We also thank Yuelong Wu from the Anhui Division of Wildlife Conservation. The Anhui Research Center for Chinese Alligator Reproduction (ARCCAR) has long been the center of conservation activities for this beleaguered species, and we are extremely grateful to its past and present directors, Wanshu Xie and Jialong Zhu, vice-director Chaolin Wang, and other staff, especially Min Shao, Renping Wang, Guangrong Ma, Hongxing Zhu, Yongkang Zhou, Xuhong Zhang, Guohong Wang, Xiuhong Li, and Xiaodong Gao.

Likewise, we developed a close and productive relationship with the Chinese alligator breeding center in Changxing, Zhejiang Province, and we would like to particularly thank its current and past directors, Weizheng Yang and Zhiping Wang, as well as Mingshui Huang, Zhenwei Wang, Yunfei Hao, Dabin Ren, Pei Zhu, Yan Zhan, Mingjuan Cheng, Huoquan Zou, and Baoshan Chen. In the Department of Biology at East China Normal University we had the support of a great group of people, and we would like to specifically extend our thanks to Professor Helin Shen, Yunmei Liang, Youzhong Ding, Lijun He, Jianshen Wu, Zhenhuan Wang, Wei Wu, Kejia Zhang, Hua Wang, and Jun Han. At Anhui Normal University, we thank Xiaobing Wu and especially Bihui Chen, who generously shared his vast experience of Chinese alligators when we met him in 1997.

Our work on Chongming Island was ably assisted by Yiming Xie, director of the Wildlife Division of the Shanghai Forestry Bureau; Enle Pei, director of the Shanghai wildlife conservation station; and at the Chongming Dontan Nature Reserve, assistant director Denchang Du, director Guoxian Song, and staff. Zhonghong Qiu and Yongjian Guan from the Shanghai Dontan Wetland Park were exceptionally helpful in many ways.

In China, we had invaluable support from the Wildlife Conservation Society (WCS) China program staff, particularly Endi Zhang, the first director of WCS-China, and subsequently Yan Xie, the current director. Both gave generously of their time to make this project possible. Other WCS staff in China who made our work possible are Aili Kang, Eve Li, Lishu Li, Tanya Tang, Zhirong Li, and Xiang Zhou. In recent years, much of the responsibility for WCS activities with Chinese alligators has been taken over by Shunqing Lu and Zhang Fang.

The Bronx Zoo, WCS's flagship zoological park, has had a long tradition of support for Chinese alligator conservation. Peter Brazaitis, then collections manager for the reptile department, launched the first attempts to breed the species in the Bronx in the 1960s. His wife, Myrna Watanabe, was the first Western scientist to work on Chinese alligators in modern-day China. Both provided us with encouragement for our field studies. John Behler, former curator of reptiles, led early efforts to breed the species in captivity by collaborating with other U.S. zoological parks and moving their breeding stock to the more agreeable environment of the coastal marshes of Louisiana. Following the first successful breeding of the species in 1977, John coordinated the American Zoo and Aquarium Association (AZA) Chinese alligator program for many years. John worked with us in China in 2001 and was an indefatigable herpetologist and good friend. Sadly, John, who was always so full of life, died in February 2006 at the early age of 62 and left a huge void in the field of reptile conservation.

The Asia regional program of WCS was very supportive of our work, particularly directors Josh Ginsberg and Colin Poole and assistant director Peter Zahler. Rose King efficiently handled many of our logistical needs. The WCS's senior vice president for global conservation, John Robinson, offered unending encouragement for our endeavors with Chinese alligators. On the technical side, Bonnie Raphael, WCS veterinarian, provided expertise in China in selecting healthy animals for reintroduction and attaching radio transmitters to their tails prior to release. George Amato, now director of the Sackler Institute for Comparative Genomics at the American Museum of Natural History, was extremely helpful in his collaboration with us and with Xiaobing Wu from Anhui Normal University to conduct genetic evaluations of Chinese alligators. Eleanor Briggs, an unflagging supporter of WCS conservation activities in Asia, helped us document the first release of alligators in Hongxing.

As a graduate student finishing up at the University of Florida, one of us (JT) was hired by the International Union for Conservation of Nature Crocodile Specialist Group to compile the first Action Plan for the conservation of crocodilians worldwide. It was during this process that he became concerned about the paucity of information on the Chinese alligator. After being hired by the Wildlife Conservation Society in 1993, JT began making contacts through WCS's far-flung network of field staff to see if there were any Chinese groups interested in collaborating on a Chinese alligator research and conservation project. A first attempt to work with Chinese alligators in 1993 failed, and several years passed with JT becoming occupied with other projects.

It was George Schaller, the renowned WCS field biologist, who finally catalyzed our collaboration. George had been working on giant pandas in the 1980s when he met Xiaoming Wang, then a graduate student at Sichuan University, in the forests of the Wolong Reserve in western Sichuan Province. When XW finished his PhD in France and returned to China, they continued to collaborate on a variety of projects as XW advanced in his career and became a professor at the East China Normal University in Shanghai. It was a cold winter day in 1996 in the basement of the education building at the Bronx Zoo, the old headquarters of the WCS International Programs, when George first suggested we collaborate. George had just received the prestigious International Cosmos Prize, an award given in recognition of his achievements in the field of wildlife conservation. The prize came with some money, which George dedicated to a variety of wildlife projects, one of which was the purchase of a new field vehicle for XW. The sites where alligators were reported to live were less than a day's drive from Shanghai, and George suggested we use this car to carry out a collaborative project. Thus our work began.

None of our efforts would have been possible without the generous financial support of a number of organizations. Aside from the considerable investment made by the Wildlife Conservation Society, much of our initial work was funded by the Walt Disney Company Foundation, the Conservation Endowment Fund of the AZA, the AZA Crocodilian Advisory Group, and the St. Augustine Alligator Farm Zoological Park. Kent Vliet (University of Florida), Andy Odum (Toledo Zoo), John Brueggen (St. Augustine Alligator Farm Zoological Park), and Lonnie MacCaskill (Disney) all played key roles in helping to secure this funding. Since 2007, work on Chinese alligators has also been generously supported by the Ocean Park Conservation Foundation in Hong Kong. Support in China was provided by the East China Normal University, the National Science Foundation of China, the Key Disciplines Program of Shanghai, and the Ministry of Education of China.

The IUCN Crocodile Specialist Group (CSG) has for many years played a key role in the international coordination of crocodilian conservation efforts, and we are thankful to Harry Messel, Graham Webb, Perran Ross, and Adam Britton for their long-term support of programs to conserve the Chinese alligator. F. Wayne King, first director of conservation at the Bronx Zoo and chair of the CSG for many years, provided steadfast support for JT throughout his graduate work at the University of Florida.

The writing of this book was greatly assisted by many people, most of whom we have already thanked above. Kraig Adler provided us with bibliographic information on Albert Fauvel and other herpetologists. Chris Brochu helped us understand the fossil past of alligators and the debate about how, and when, they may have arrived in Asia. During our first full survey of sites with alligators in Anhui Province in 1998, we were accompanied by Scott McMurray, who provided us with camaraderie and expert advice on the potential effects of agrochemicals on alligators and other fauna in the rice fields of Anhui. Lisa Ortuno (née Davis) and Jinyung Fang generously allowed us to use figures from their publications. The renowned painter Peng Ye from East China Normal University created a series of paintings that depict the relationship between people and Chinese alligators and generously donated them for use in the book. Kent Vliet graciously shared his encyclopedic knowledge of crocodilian behavior and the biology of American alligators. To all, and to those we have inadvertently failed to mention, we owe a full measure of appreciation.

Abbreviations

AFB	Anhui Forestry Bureau
ARCCAR	Anhui Research Center for Chinese Alligator Reproduction
AZA	Association of Zoos and Aquariums
BP	before the present
CABC	Changxing Alligator Breeding Center
CITES	Convention on International Trade in Endangered Species of Wild Fauna and Flora
ECNU	East China Normal University
HOTA	head oblique, tail arched
IUCN	International Union for Conservation of Nature
LAG	lines of arrested growth
MCZ	Museum of Comparative Zoology
MVP	minimum viable population
mya	million years ago
NCAR	National Chinese Alligator Reserve
NGO	nongovernmental organization
PRC	People's Republic of China
PVA	population viability analysis
RAPD	randomly amplified polymorphic DNA
SAV	subaudible vibrations
SEPA	State Environmental Protection Administration
SFA	State Forestry Administration
SIIC	Shanghai Industrial Investment Company
SSAR	Society for the Study of Amphibians and Reptiles
SSC	Species Survival Commission
SSP	species survival plan
SVL	snout-to-vent (body) length
TL	total length
TSD	temperature-dependent sex determination
UNDP	United Nations Development Program
WCS	Wildlife Conservation Society
WWF	World Wildlife Fund

The Chinese Alligator

The Earth Dragon

When most people think of alligators they think of the relatively large, broad-snouted denizens of the swamps and marshes of the southern United States. The American alligator is well known in popular culture and is also the most thoroughly studied of the world's crocodilians. It comes as quite a surprise to many that there is also a "Chinese" alligator. Although the American alligator grows to a much larger size than the Chinese species, the two share many behavioral and ecological features. But mirroring their disparate distributions on different sides of the planet is the wide gulf that exists between the population status of the two species: the American alligator being one of the world's most abundant crocodilians and the Chinese alligator, the rarest.

The alligator has both a modern and an ancient history in China. Much of its earliest history is entwined with that of the Chinese dragon, and in Chapter 4 we discuss the role of the alligator in the development of the Eastern dragon mythology. The modern story begins in the mid-nineteenth century with the scientific description of the Chinese alligator by the French naturalist Albert-Auguste Fauvel (Chapter 3). But notwithstanding the long history of human interactions with the Chinese alligator, relatively little was known about its biology until quite recently, in part due to its rarity. Within recent historical times the alligator has never been abundant. Despite extensive searching, Fauvel was only able to locate a small number of specimens for his report on the species in 1879. As a result of continued habitat loss and uncontrolled killing of animals, even as early as the 1920s some experts suggested that the Chinese alligator was extinct (Werner 1928). The truth was that while populations continued to decline, scattered groups of alligators remained, principally in southeastern Anhui Province and parts of adjacent Zhejiang Province. Following the Communist Revolution in 1949, studies of alligators were briefly undertaken by Chinese scientists before being halted during the periods of social upheaval known today as the Great Leap Forward (1958–1962) and the Cultural Revolution (1966–1976). During that time the lack of up-to-date information led some Western authorities to once again speculate that the Chinese alligator might have become extinct in the wild or at least that its numbers had diminished past the point of no return (Neill 1971; Minton and Minton 1973; Inskipp and Wells 1979). In the early 1980s Chinese and U.S. biologists collaborated on a study of the alligator's status and ecology. Most of what we know about the ecology and behavior of the species is the result of work initiated since the 1980s on captive animals, for by that time the wild population had all but disappeared.

Figure 1.1. An adult male Chinese alligator (*Alligator sinensis*), referred to in parts of southern Anhui Province as *tu long*, the earth dragon.

As China recovered from the effects of the decade-long Cultural Revolution, environmental conservation issues were again addressed. Efforts to protect the alligator began in 1972, when the Chinese government listed it as a Class 1 endangered species, providing it with the highest degree of legal protection (Z. Wan et al. 1998). In 1979 the Fifth National People's Congress passed a new forestry law (which included wildlife conservation), and the first ever PRC meeting to designate protected areas was held in 1981. In December 1982 the new national constitution stipulated that "the State ensures the rational use of natural resources and protects rare animals and plants" (W. Li and Zhao 1989). Nevertheless, regulations prohibiting the capture or killing of alligators were not entirely effective, particularly during the 1970s and early 1980s (Watanabe 1981). Also, from the beginning, conservation efforts for alligators focused almost entirely on captive breeding, with programs established in Anhui and Zhejiang provinces in 1979. To create the breeding centers, many of the remaining wild alligators were caught in the early 1980s. This practice continued almost to the present day, largely through the collection of eggs from the ever-dwindling number of wild nests.

In the 1980s and 1990s the Chinese alligator was internationally acknowledged to be a critically endangered species. The American alligator, which only a few decades earlier had been endangered, recovered to such a remarkable degree that it is once again a common resident of wetlands in the southeastern United States. Today, American alligators are an internationally acclaimed wildlife conservation success story (Chapter 2). Surely the Chinese alligator, a much more innocuous species due to its small size and its predilection for spending much of the year hiding in underground burrows, could stage a comeback. In fact, before we began our work, the news from China was very encouraging. In addition to the alligator captive breeding program, the Chinese government had established the National Chinese Alligator Reserve, a relatively large region where the last remaining groups of alligators were found. Once husbandry protocols were established, the captive breeding program took off and the number of captive alligators soared (Chapter 8).

As a result of the proliferation of captive animals, in 1992 the Chinese government was authorized by the CITES, the UN body that regulates trade in endangered species, to initiate commercial export of alligators and their products from the Anhui breeding center. The rationale was to promote international trade and generate income for the operation of the breeding center as well as other initiatives that would provide economic and conservation benefits for the management of wetlands and alligators (PRC 1992). However, it soon became clear that there was no market for skins and only a minuscule one for the sale of live alligators to zoos or pet collectors. The one potentially large market, domestic sale of alligator meat, was tightly controlled by the national government. In order to generate income to support the captive breeding program, the Anhui center was forced to reinvent itself as a local zoo and amusement attraction.

While the captive breeding programs were being set up in the 1980s, little information was available on what was happening with the wild population. This changed in the early 1990s when, as a result of the CITES application for registration of the Anhui breeding center, the Chinese government was required to provide information on the status of alligators in the wild. The report was promising, indicating that approximately 800 alligators remained in the wild in Anhui and parts of neighboring Zhejiang Province and that the wild population was increasing, with a dozen nests found in recent years and hatchling survival rates of 15% to 30%. Additionally, as part of the process to evaluate China's CITES application, a review of the Chinese alligator conservation program was carried out by the IUCN's Crocodile Specialist Group (Webb and Vernon 1992). The findings of this review were even more encouraging, with overall population estimates of some 1,000 wild alligators in Anhui Province, 900 of which were found within the NCAR. Moreover, the report confirmed that wild populations were rebounding rapidly, at a rate of up to 15% per year in some areas. Unfortunately, the review was carried out during the winter (when the alligators were hibernating in their burrows), and the information compiled from various sources was impossible to verify. Nevertheless, these two reports gave a very optimistic view for the future of Chinese alligators.

It was with this promising outlook that we began our work in 1997. In July we made a preliminary trip to Anhui Province, unsure of what to expect as we drove west from Shanghai through the increasingly rural landscape. At first glance the region seemed to hold promise; the roads were flanked by ponds and verdant flooded fields. But once we had the opportunity to walk through some of these areas our impression quickly changed. Around the ponds, virtually every square meter was under cultivation, mainly for rice. The premium on land was clear when we found it difficult to skirt the edge of the ponds without trampling crops planted right up to the water's edge. Discarded packets of pesticides and herbicides littered the narrow paths we followed. Although ponds dotted the countryside, they were being used for many purposes: for watering crops, as bathing holes for water buffalos, and for raising fish or domestic ducks. The absence of birds lent a bleak and empty feeling to the landscape.

Alligators, like many crocodilians, are difficult to see during the day, when they hide in burrows or underwater. They become more active at night and can be located by their reflected eyeshine. Our first night survey was planned for Shaungken, one of

13 sites that had been officially designated for the conservation of the Chinese alligator within the National Chinese Alligator Reserve. The 1992 IUCN report indicated that there were 82 alligators at Shaungken and that the wild population in this area was growing at an annual rate of 15%. Visiting the site five years later, we were excited by the possibility of finding in excess of 100 animals. Leaving the Anhui breeding center in the afternoon, we drove west along a road that was in the process of being rebuilt. The entire route was one long obstacle course of heavy machinery, cars, and people walking, bicycling, or tending buffalos. Then we took a narrow, slippery dirt road atop a newly constructed dike that rose high above the valley floor. The dike was part of a canal system to divert water from a reservoir to downstream areas, as well as to control floods in the area around Shaungken. Neither of us had ever seen a Chinese alligator in the wild and we were filled with anticipation as we walked down a slope cultivated with pine trees to reach a series of small ponds amid the rice paddies. Night had fallen and we strapped on our headlamps and started walking through crops that extended down to the margin of the water. We quickly spotted our first alligator, but after several hours our excitement subsided. We had counted only 3 animals, and the local farmers said there were no more than 20 alligators in the ponds (and in 1999 we found even this figure to be overly optimistic).

After a few days, we visited Hongxin, another site that was reported to be among the best in the NCAR. A much larger pond (ca. 8 ha) than any we had seen at Shaungken, it had a small island. We were on a tight schedule and were forced to carry out this survey under less than ideal circumstances (in the midst of typhoon Winnie, which had devastated the southern coast of China the previous week). A local farmer reluctantly took us out in his small rowboat, the only one on the pond. Standing and leaning forward in an effort to propel the boat against the wind, our boatman unintentionally served as a sail. We were quickly pushed to the far side of the lake and were forced to return on foot. By the time we arrived back at the farmhouse, we had seen only one alligator. While the conditions were not conducive for counting alligators, a farmer who had lived at Hongxin for 70 years told us that there were no more than 10 animals in the pond. Every year since 1982 he had found an alligator nest on a small island in the pond. But each year the staff from the alligator breeding center collected the eggs, and they had never returned any of the young. The following day we visited a third site, Zhuangtou, where we saw the remains of two alligator nests from which eggs had recently been collected and taken to the already overcrowded alligator farm.

By the end of the 1997 visit our initial optimism had faded. At first we were incredulous that the situation we found at these sites could be so different from the one painted by the previous reports. Sites at which we expected to find thriving populations of alligators were found to have only a handful. Nevertheless we remained hopeful that at other sites in the NCAR we would find areas of good habitat with significant numbers of alligators. The collection of eggs from the few wild nests was worrisome, but we had no clear idea at the time just how injurious this was to the wild population. As a result of this preliminary trip we reached an agreement with the Anhui Forestry Bureau to carry out a systematic survey of all the wild alligator sites in Anhui Province. In July 1998 we met in Singapore to attend a regional meeting of the IUCN Crocodile Specialist Group, which included Chinese wildlife managers from the State Forestry Administration (Beijing) and the AFD. We held a series of productive discussions about the development of a collaborative program, and a paper was presented by Ziming Wan summarizing what was known about the status of Chinese alligators and proposing conservation actions (Z. Wan et al. 1998).

From Singapore we went to the Anhui breeding center, officially known as the Anhui Research Center for Chinese Alligator Reproduction. There we finalized details of the survey that we would carry out with two students from East China Normal University and two biologists from the AFB. Our first stop was Zhuangtou, one of the NCAR's designated alligator sites. We were in a Jeep Cherokee donated to ECNU by George Schaller, the renowned WCS scientist who had led pioneering studies of giant pandas in China in the 1980s, and a "Beijing jeep" from the AFB. In this two-car convoy we passed through the city of Xuancheng, where the ARCCAR is located, and then followed a series of narrow, rural roads toward our destination. Driving in China can be an unnerving experience, particularly for Westerners who are not familiar with the set of unwritten rules that direct, in a loose sense, vehicular navigation on Chinese roads. In recent years the Chinese government has made great strides in improving adherence to driving conventions, but driving can still be a harrowing experience. In 1998 in the Anhui countryside, the narrow roads were clogged with bustling foot traffic, bicycles, buffalos, tractors, cars, and trucks, none of which wanted to yield to anyone else; a cacophony of blowing horns, shouting people, and bellowing animals filled the air. Driving from site to site resembled a competitive slalom, constantly weaving around obstacles in the road at speeds that seemed 50% faster than were appropriate. Our driver, from ECNU, was an expert at maneuvering on the roads, but the ride was still unsettling.

On that first day of our 1998 survey we were following behind the AFB jeep, with its police siren and bullhorn. Coming in the other direction was a car driven by a distracted police officer from a neighboring county. To get around a truck parked on the road, the driver swerved out into the middle of the road just after our lead car had passed. The driver did not see us following close behind and the result was a head-on collision. Both of us were sitting in the backseat of our Jeep, along with Changming Gu from AFB. JT was the unluckiest, with a broken femur, a broken shoulder, and a deep gash in his scalp. The police officer in the other car was in worse shape but survived to spend time in jail for precipitating the accident. After getting to know hospitals in Xuancheng, Shanghai, Hong Kong, and finally New York, JT made a full recovery, but our plans for alligator surveys were put on hold. Perhaps the accident was an omen, as 1998 was also the first year, most likely in the entirely evolutionary history of the species, that not one Chinese alligator nest was found in the wild.

In 1999 we returned with freshly installed seatbelts in the rear seats of the vehicle and firm instructions for the (new) driver. We also had a new student, Youzhong Ding, and another team member, Scott McMurray, an environmental toxicologist from Texas Tech University who was helping us evaluate the impacts of agrochemicals on alligators. That year heavy rains and flooding nearly thwarted our survey plans. During our first week we settled into a routine of driving to a new site each morning, having lunch with a group of local dignitaries, who always went out of their way to treat us as special guests, and in the afternoon carrying out daytime reconnaissance of nearby sites. We would walk around each site, looking for signs of alligators, footprints or tail drags in the mud, burrows, or brief glimpses of alligators themselves, and interview local residents. If we found even the slightest evidence that alligators had been in the area in the last three years we would remain until after dark and search for eyeshines in the beams of our flashlights. The ponds were usually small, and in most areas we could easily walk around them following narrow paths atop water control banks. In a few areas that had larger ponds we used local boats, but we were also prepared with a small inflatable boat that could (barely) seat two people. Our survey team, with its unusual objective and with two foreigners, the first seen by most people at nearly all the sites we visited, invariably attracted a great deal of attention wherever we went. But it was our inflatable boat that added greatly to local interest and that, at times, resulted in a long line of people, old and young alike, following us along the narrow paths through the rice fields as we carried our little boat from pond to pond.

We had planned to start our 1999 surveys in Langxi County, one of five counties in the NCAR. However, when we arrived in July surveys in Langxi were not possible because much of the area was under water following the rupture of a major levee. We completed surveys in the other four counties, hoping that the waters would subside enough for us to visit Langxi. When we finally did get there we found the city still awash in mud and debris. The Langxi representative for the Anhui Forestry Bureau, whose responsibility it was to guide us to the sites in his county, could not be found. At the AFB office we were told he was trying to salvage what he could from his house, which had been flooded to the second floor. Eventually, we were able to visit the three sites in Langxi, only to find that two sites had not had any alligators for over 10 years and at the third site people were up in arms because the last remaining alligators had been eating their ducks.

The bleakness of the situation had become apparent to us even before our visit to Langxi. For one month we had traveled from county to county visiting all the sites where alligators were known to be found. Each day we hoped that at the next site we would find a sizable area of natural habitat with a healthy alligator population. But it never happened. Site after site was either small and heavily used by people in adjacent villages or biologically ill-suited for alligators. The day we arrived at Changle, a small farming village in Nanling County, was particularly hot, and the air was thick with smoke as farmers were burning their fields in preparation for planting their second crop of rice. We arrived at the site by car, driving up right to the edge of the tiny pond and stepping out into the patio of the site's caretaker. With our students from the ECNU and representatives from the Anhui Forestry Bureau we walked around the pond (which if we had not stopped to look carefully for signs of alligators would have taken us all of 10 minutes) and interviewed some of the local farmers. The caretaker, paid a monthly stipend by the AFB to protect the alligators, started calling them by clapping his hands and making a series of guttural calls. When no alligators appeared he explained that throwing food into the pond was the best way to see the alligators, but as we had not brought any suitable bait he would have to make do with something else. He then bent down and picked up the charcoal remnants of his cooking fire and tossed these into the pond. No alligators responded to this either. Before nightfall we did see one alligator. Later, a farmer showed us a photograph of an emaciated alligator that had been living in the pond but that had abandoned it the previous spring and had been found wandering through a nearby drainage canal, apparently looking for a better place to live. Based on the large size of the animal it

Figure 1.2. Surveying for Chinese alligators at Changle village, Nanling County, in southern Anhui Province.

was almost certainly a male that had been captured, photographed, and returned to the pond. The only known adult female at this site had been found dead in 1997, her belly full of mice that had been poisoned in nearby houses and tossed into the pond. The remaining animals, we counted four that night, were survivors of a nest that in 1996 had escaped collection by the staff of the ARCCAR.

While waiting for nightfall and the chance to count alligators, we climbed up to the top of a local house overlooking the site. Emerging from the narrow stairway and onto the flat roof we pushed our way past a maze of clothes hanging out to dry and over to the edge of the roof, where we had a clear view of the surrounding landscape. The small pond lay before us, surrounded by a sea of rice paddies, hemmed in by a road, a cemetery, and two other houses. A stream of people were riding their bicycles or walking along the road by the caretaker's house, and on the other side of the pond a farmer was using a mechanical tiller to plow the soil up to the water's edge. What made all this particularly disheartening was that in the previous weeks we had been eagerly looking forward to visiting this site as it was reported to contain the largest remaining group of Chinese alligators.

The Changle visit, coming as it did near the end of this frustrating and fruitless search for a "good" alligator site, underscored the depth of the problems facing the wild alligators and belied the previous optimistic reports on the status of the wild populations. In the midst of the Chinese government's efforts to build up a captive population of Chinese alligators, the fate of the last wild alligators was being ignored. The small and fragmented nature of the population was alarming. Our results showed that no group of "wild" alligators had more than two adult females; some sites appeared to have just one individual. Of the 13 sites in the NCAR officially designated for the conservation of alligators, 3 had no alligators whatsoever. The last remaining wild alligators were living out their years in sites ranging from tiny ponds in the midst of villages to small, biologically unproductive impoundments in the hills surrounding the agricultural valleys. Most of the "habitat" was being used by local villages for agriculture or aquaculture. Nesting was down to minimal levels, and many, if not most of the eggs being laid each year were collected and brought to the already overcrowded Anhui breeding center.

The future of the Chinese alligator looked grim. How could we even begin to propose conservation measures for alligators when human demands on the landscape were so intense? Everywhere we looked there were poor farming communities struggling to make a living off the limited amount of available land. The ponds used by alligators were important to people for rearing fish or ducks, watering the family buffalo, and washing the laundry. Marshes and shallow areas fringing ponds, the typical habitat for juvenile alligators, were prime sites for planting rice. Alligators were considered by many to be troublesome creatures because they burrowed into the complex array of dikes that control water movement through the rice paddies and ate ducks and fish. Most of these areas had been so altered that we actually found the breeding ponds in the ARCCAR to be the best alligator "habitat" that we had seen during our survey.

Nevertheless, the surveys did have their high points. At some sites people showed a great deal of local pride at having alligators in their ponds. One of our best moments came exactly one year to the day after our car accident. On 21 July 1999 we found our first active Chinese alligator nest on a ridge in a pine forest above the ponds at Zhuangtou, For us this was an incredibly uplifting experience. Despite the sad state of the wild population, this nest was, for us, a symbol of the resilience and adaptability of the Chinese alligator and gave us hope that, if given the chance, the alligator could return to areas that were adequately protected.

Today, the conservation of wild populations of Chinese alligators represents a unique challenge, for it is literally a species without a habitat. Our feelings of despair as we stood on the roof in Changle have since been tempered by the realization that the Chinese alligator is a remarkably adaptable creature that can thrive given even minimal living conditions. The captive breeding program has been successful and there are now more than 10,000 living Chinese alligators. On the other hand the more difficult issues of habitat conservation and working proactively with local communities to promote the species' conservation in the wild have been almost entirely avoided by Chinese wildlife authorities. Given the enormous human population pressures on the alligator's habitat and the lack of resources available for wildlife conservation in China, one can almost sympathize with the decision to take this path of inaction. Nevertheless abandoning the wild populations is, in our opinion, the wrong path to take, and despite the difficult situation there are ample opportunities to guarantee the long-term conservation of Chinese alligators in the wild.

This book describes what we have learned about Chinese alligators over a period of more than a decade. We first place the plight of the Chinese alligator in a global context by providing an overview of crocodilians and their conservation (Chapter 2) and in a historical context by describing the long and fascinating history of alligators in China (Chapter 3). The historical context is important for understanding the cultural links between the alligator and its better known relative the Chinese dragon (Chapter 4). In Chapter 5 we outline what is known about the biology of Chinese alligators and make comparisons to the much better studied American alligator. The long history of the decline of the Chinese alligator, paralleling the growth of the lower Yangtze as China's breadbasket and the loss of wetlands, is described in Chapter 6, which introduces the discussion of the present state of alligator habitat and populations in Chapter 7.

Our principal reason for writing this book was to promote conservation programs to restore wild groups of alligators to parts of its former range. The final chapter outlines the conservation challenges faced by the alligator, and our vision for the species' future.

2

The Family of Dragons

Crocodilians of the World
and Their Conservation

*The nearest thing to a dinosaur that anyone can see alive in the wild today
is an alligator or crocodile, but one had better not postpone the opportunity
too long. These big reptiles are rapidly going the way of the dinosaur.*

—S. A. Minton and M. R. Minton, *Giant Reptiles*, 1973

AN INTRODUCTION TO CROCODILIANS

Crocodilians are reptiles, a group that by popular notion includes turtles, snakes, and lizards, and the unusual lizardlike species of tuataras found only in New Zealand. Unlike the other reptiles, they are descended from the ancient stock that also gave rise to birds. Although birds (Aves) are quite different in terms of ecology and behavior, they are included in the class Reptilia due to their close evolutionary links with the crocodilians. Together, birds and crocodilians are the living representatives of the Archosauria, or "ruling reptiles," that include the now-extinct dinosaurs and flying pterosaurs. The modern crocodilians (Eusuchia) have a fossil history that can be traced back to the Mesozoic, more than 65 mya. Most crocodilians alive today, however, are of fairly recent evolutionary origin and are classified into three families: the Alligatoridae (alligators and caiman), the Crocodylidae (crocodiles), and the Gavialidae (gharials). Currently there are 23 species of living crocodilians found mostly in the tropics and subtropics (Table 2.1).

The crocodiles are the most diverse group of crocodilians, accounting for more than half of all living species. The true crocodiles (genus *Crocodylus*) are composed of 11 species found in Asia, Africa, and the New World, and appear to be a relatively recent evolutionary radiation (Brochu 2000; McAliley et al. 2006). Two African species in the family Crocodylidae, the African dwarf crocodile (*Osteolaemus*) and the African slender-snouted crocodile (*Mecistops*) are in separate genera. The two gharials are very large, slender-snouted Asiatic species. The evolutionary history of these two species is controversial, with morphological characters linking the Malayan

Table 2.1   The living crocodilians

Common name	Scientific name	Distribution	Maximum size (m)	Status classification
American alligator	Alligator mississippiensis	SE United States	4	NT
Chinese alligator	A. sinensis	E China	2	CR
Spectacled caiman	Caiman crocodilus	South and Mesoamerica	2.5	NT
Yacare caiman	Ca. yacare	Central South America	3	NT
Broad-snouted caiman	Ca. latirostris	Central South America	3.5	NT
Black caiman	Melanosuchus niger	Amazon basin	5.5	LC
Dwarf caiman	Paleosuchus palpebrosus	Northern and central South America	1.5	NT
Smooth-fronted caiman	P. trigonatus	Northern and central South America	2	NT
American crocodile	Crocodylus acutus	Northern Neotropics	5	VU
Morelet's crocodile	C. moreletii	Northern Caribbean Mesoamerica	4	LC
Cuban crocodile	C. rhombifer	Cuba	3.5	CR
Orinoco crocodile	C. intermedius	Orinoco basin	6	CR
Nile crocodile	C. niloticus	Africa	5.5	NT
Mugger crocodile	C. palustris	Indian subcontinent	4	VU
Estuarine crocodile	C. porosus	Indomalaya–SE Asia and Oceania	6.5	NT
Siamese crocodile	C. siamensis	SE Asia	4	CR
Philippine crocodile	C. mindorensis	Philippines	3.5	CR
New Guinea freshwater crocodile	C. novaegineae	New Guinea	3.5	NT
Australian freshwater crocodile	C. johnsoni	Australia	3	NT
African slender-snouted crocodile	Mecistops cataphractus	Central and West Africa	4	DD
African dwarf crocodile	Osteolaemus tetraspis	Central and West Africa	1.8	VU
Malayan gharial	Tomistoma schlegelii	Indomalaya	5.5	EN
Gharial	Gavialis gangeticus	Indian subcontinent	6.5	CR

Note: Status classification is based on the IUCN Red List: NT = not threatened; LC = least concern; VU = vulnerable; E = endangered; CR = critically endangered; DD = data deficient.

gharial more closely with the crocodiles, but genetic analyses showing it to be close to the gharial (Gatesy et al. 2003).

The alligator family has two distinct groupings, a tropical group of six species referred to as caiman, and two true alligators (the genus *Alligator*) the American alligator and the Chinese alligator, which are the only species to inhabit areas within the subtropical-temperate climatic transition. The Chinese alligator is the only living species in the family Alligatoridae that is found outside of the New World.

Alligators initially came to the attention of the Western world when the first Europeans arrived in North America in the sixteenth and seventeenth centuries. Landing with preconceived notions about the world around them based on their knowledge of the natural history of the Old World, they encountered a new and wondrous set of plants and animals. The large, amphibious reptiles seen in Florida and other parts of the southeastern United States were generally thought by most educated people to be crocodiles, which were well known from the descriptions of Nile crocodiles dating back to the Greek historian Herodotus. Nevertheless, it was the anglicization of the Spanish word *el lagarto* (the lizard) that came to be popularly applied to these animals, now known as the American alligator. The term "alligator" became so ingrained in the English language that after the discovery of the New World, English colonists who settled in places where there are not, and never have been, true alligators applied the word "alligator" to true crocodiles found in other parts of the world, a practice that is reflected in place-names such as Australia's Alligator River. Unlike its better-known American relative, the Chinese alligator remained unknown to Western scientists until the nineteenth century.

CROCODILIAN CONSERVATION AROUND THE WORLD

As large predatory animals, crocodilians have long held an ambivalent place in the hearts and minds of people (Chapter 4). Like many other types of wildlife, crocodilians were used as a source of food by many human cultures (Klemens and Thorbjarnarson 1995). The prehistoric extinction of some island forms of crocodiles, such as the unique terrestrial crocodiles (*Mekosuchus* sp.) from Fiji, Vanuatu, and New Caledonia (Mead et al. 2002; Molnar et al. 2002), may be linked to humans' hunting them for food or killing them out of fear. But it was their skin, not their meat, that created the first modern conservation crisis for crocodiles and alligators.

Evolution has endowed crocodilians with a remarkable collection of adaptations for life as the top aquatic predators in tropical wetlands. Over the last two centuries, however, some of these adaptations have become a double-edged sword. Their tough but pliable hide has been much sought after for the manufacture of products ranging from shoes and bags to wallets and belts. The reflective tapetum lucidum, a layer of guanine crystals behind the retina in their eyes, so useful for amplifying the dim light available at night and improving their nocturnal vision, made them an easy target for nocturnal hunters with a flashlight. Beginning in the nineteenth century and extending through the 1960s, the use of leather made from the skins of alligators and crocodiles was fashionable. Many species were locally abundant and were easy to find during annual periods of low water level, facilitating large-scale hunting. During this time, the commercial demand for crocodilian leather was the primary threat to their survival; populations around the globe plummeted. In the 1960s and 1970s there was a growing realization that many crocodilians were becoming rare as a result of commercial hunting. This was followed by a transition period in the 1970s and 1980s with worldwide efforts to control the trade of skins and the emergence of numerous programs that produced hides on a controlled, regulated basis. We have now entered a second era, that of managed use and declining habitat. Today, many of the species that were rare in the past as a result of unregulated hunting have recovered (e.g., the American alligator), and habitat loss is looming as the principal conservation issue for a well-defined group of crocodilians.

The first known period of commercial skin hunting for American alligators was during the early nineteenth century in Louisiana, ending some time before 1827. The famed naturalist John James Audubon, who at that time was living in the lower Mississippi River valley, commented on how unfettered hunting had reduced the number of alligators in the area (Kellogg 1929). While hunting appears to have been locally intense, the limits of tanning technology at that time apparently kept the leather

from being particularly useful as footwear. Commenting on their former abundance in the Red River, Audubon noted:

> It was on that river particularly that thousands of the largest sizes were killed, when the mania of having either shoes, boots, or saddle-seats, made of their hides, lasted. It had become an article of trade, and many of the squatters and strolling Indians followed for a time no other business. The discovery that the skins are not sufficiently firm and close-grained, to prevent water or dampness long, put a stop to their general

Figure 2.1. (*facing page and above*) Three crocodilians: (a) the Indian gharial (*Gavialis gangeticus*), the most slender-snouted of the crocodilians; (b) the American alligator (*A. mississippiensis*); and (c) the American crocodile (*Crocodylus acutus*).

destruction, which had already become very apparent. The leather prepared from these skins was handsome and very pliant, exhibiting all the regular lozenges of the scales, and able to receive the highest degree of polish and finishing.

Alligator skin hunting increased again around 1855, when "some thousands of hides" were taken in Louisiana (Reese 1915). However, the use of alligator skin leather products really began to come into its own following the end of the Civil War. Starting about 1865, alligator skins were tanned in New Orleans and sold to clients, mostly in New York, including the wife of the famous American railroad magnate Cornelius Vanderbilt, who had a chair upholstered with alligator hide. She apparently was so enchanted with the chair that she ordered more skins, thus helping create the beginnings of a worldwide industry (Glasgow 1991).

The demand for alligator leather products soon spread to Europe, particularly France, and by 1869 more and more alligators were being hunted in Louisiana to supply the raw material. The skins were used for a variety of purposes. During the 1880s alligator skin saddles became fashionable in Russia, because the scales supposedly kept the body from sliding in the saddle. In addition to shoes, saddles, and the occasional piece of furniture, alligator skins were used for purses and handbags, boxes for player piano rolls, belts, saddlebags, card cases, slippers, trunks, traveling bags, pocketbooks, book bindings, and even petticoats.

While the earliest commercial hunting was limited to Louisiana, the great demand caused hunting to extend beyond the state and eventually even national boundaries. By the early 1890s alligator hunting had spread to Florida, which soon surpassed Louisiana in the production of skins (Glasgow 1991). These two states appear to have

Figure 2.2. Alligator hunting in Florida in the 1880s. Photo courtesy State Archives of Florida, Florida Department of State, Tallahassee.

been the only ones supplying alligator skins through the first few years of the twentieth century.

The scale of the harvest was immense. H. M. Smith (1893) estimated that between 1880 and 1893 some 2.5 million alligators had been killed. In Louisiana, McIlhenny gave a figure of 3 to 3.5 million taken between 1880 and 1933. By 1890 the effects of overharvesting were evident in Florida. Smith (1893) comments on the fact that alligators were becoming harder and harder to find throughout parts of Florida where only a few years before they had been abundant. The exotic leather fashion swept into Europe by the late 1800s, and to meet the growing demand, crocodile hunters expanded into Mexico, Central America, and the Caribbean. By the 1930s many of the skins tanned in Europe came from South America. Following World War II and the rebuilding of the European tanning industry, the demand for skins redoubled. Hunting resumed in the neotropics and spread into Africa, Asia, and Australia. For many species, hunting was intense and quickly resulted in the depletion of wild populations. From 1950 to 1965, 7.5 million caiman skins were exported from Amazonas State in Brazil (N. Smith 1981). By 1980 Medem (1981) calculated a minimum of 11.65 million caiman skins had been exported from Colombia. In South America the annual number of caiman exploited during the 1980s was estimated to be in excess of 1 million (Jenkins and Broad 1994).

A variety of national and international restrictions on hunting and trade have been enacted over the last 30 years, and populations of once overexploited species are recovering (Thorbjarnarson 1991; Ross 1998). Beginning in the 1960s the scarcity of skins also had significant impacts on the reptile leather business as many tanneries closed, purchased illegal crocodilian skins, or switched to the hides of other reptile including sea turtles, lizards, and snakes (King 1978). With the adoption of the Convention on International Trade in Endangered Species of Wild Fauna and Flora in 1975, the first steps toward regulating wildlife trade at an international level were taken. Nevertheless, because of the high demand for skins, considerable illegal trade continued (Inskipp and Wells 1979).

Despite the considerable economic incentives to kill crocodilians, in any one region commercial hunting usually was nonproductive long before populations reached levels close to biological extinction; as a result no species has gone extinct in recent times as a result of skin hunting. But where habitat loss has been a significant factor, commercial hunting and the killing of unwanted animals has led to a crisis situation. Today, the most critically endangered crocodilians, including the Chinese alligator and the Philippine and Siamese crocodiles, are the victims of the double whammy: overhunting in the past and habitat loss in the present (Ross

1998). Where habitat loss was not a significant factor, the reduction or elimina-tion of commercial hunting initiated a phase of population recovery. While this was most dramatic in the case of the American alligator, the recovery of other species such as Morelet's crocodile, the Australian freshwater crocodile, and the Nile and saltwater crocodiles (Ross 1998) has demonstrated the benefits of increased protec-tion worldwide.

Crocodilian commercial use programs have become a profitable wildlife busi-ness worldwide that includes a wide variety of stakeholders ranging from hunters, skinners, landowners, and ranchers to skin traders, tanners, and leather manufac-turers. In Louisiana the sale of alligator meat and skins was $25 million per year in the early 1990s (Joanen et al. 1997); by 2007 sales had reached $40 to $50 million per year (Elsey and Kindler 2007). In Venezuela the peak export value (in 1990) was approximately $25 million (Thorbjarnarson and Velasco 1999). While it is dif-ficult to quantify the role of economic benefits in generating conservation incentives among local communities and landowners, in some countries crocodilian manage-ment is clearly given a higher priority because of its economic potential (Child 1987). Commercial use can also generate a more positive image of crocodilians among the general populace, and through the use of severance fees and taxes, these programs can generate funds for cash-strapped wildlife management agencies, as was the case in Venezuela.

By the late 1980s illegal trade in classic skins had been significantly reduced, largely as a result of CITES (Hutton and Webb 2002). While illegal caiman trade remained a complex problem, the universal tagging resolutions adopted by CITES in 1992 and 1994 have provided an important tool for identifying the origin of skins and for regu-lating trade. Evidence points to a significant decline in illegal caiman trade between 1992 and 2002 (Hutton and Webb 2002).

Unlike the historical patterns of commercial hunting where short-term profits were the primary objective, today's managed harvests are based on an understand-ing of the population biology of the relevant species, with sustainability as a primary objective. While a basic understanding of the effects of the harvest on wild popula-tions are available for only a few areas, harvests have been designed to minimize the negative demographic effects (in most cases by targeting juveniles or adult males), and most evidence suggests that hunting levels have been reduced to sustainable lev-els (Ross 1998). In Australia and the United States detailed population monitor-ing and ecological research programs have demonstrated that harvested crocodilian populations can continue to grow (Dutton et al. 2002; Joanen et al. 1997; Webb 2000; Stirrat et al. 2001).

The history of crocodilian managed harvest programs shows both the advantages and disadvantages of an approach based on sustainable use. The demand for croco-dile skin products made commercial management of the more common species an attractive alternative for businessmen and wildlife management authorities. The suc-cess of these approaches can be measured by the number of nations that have begun managed harvests, the global shift from illegal to mostly legal skins, the amount of research and population monitoring of commercially managed species, and the pop-ulation recovery of a variety of crocodilian species (Ross 1998; Webb 2000; Elsey and Kindler 2007).

The limitations of commercial use for management of crocodilians are evident in the lack of effectiveness this approach has shown in dealing with the most highly endangered crocodilians, particularly where habitat loss has been a major contributing factor (e.g., the Chinese alligator and the Philippine crocodile). For the most part, the focus on sustainable use has had the perverse effect of shifting the attention of crocodilian managers away from highly threatened species, such as the Chinese alligator, to the more common ones with high commercial value, such as the American alligator.

THE AMERICAN ALLIGATOR: A CONSERVATION SUCCESS STORY

While the two species of alligators are much more closely related to one another than to any of the other crocodilians, their conservation status could not be more different. When the first Europeans visited the southeastern United States, they were impressed by the abundance of alligators in the rivers and lakes. As a result of extensive commercial skin hunting, population numbers plummeted in the late nineteenth and the early twentieth centuries. By the early 1960s it was evident that hunting had had a major impact on wild populations of alligators, and protective measures were tightened. In Louisiana alligators over 5 feet long were protected, and a 60-day hunting season was established in 1960. Hunting was totally closed in 1962 (Joanen et al. 1997). In Florida the hunting season was eliminated in 1961, but illegal hunting continued, and a minimum of 140,000 alligators were killed in Florida from 1962 to 1969 (Hines and Percival 1987). The main problem was that laws in individual states had little power to close down the interstate traffic in alligator skins. Alligators were put on the federal list of endangered species in 1967, but this also was relatively ineffective in stopping illegal hunting. The turning point in the battle against the illegal trade came in 1969 when a federal law—the Lacey Act—was amended to prohibit the transport of alligator skins from one state to another, providing law enforcement officials with their first real tool to close down the alligator skin markets in New York and other locations.

The positive effect was almost immediate. Throughout the South, alligator populations began recovering. By the mid-1970s surveys by wildlife officials in Florida and Louisiana found alligator populations were growing quickly. In Florida alone the state was getting 4,000 to 5,000 complaints a year about "nuisance" alligators around people's homes or in public areas.

Alligators had recovered to such a degree that beginning in the mid-1970s the federal government started reclassifying alligators from endangered to threatened, first in parts of Louisiana, then in 1977 in all of Florida and parts of South Carolina, Georgia, and Texas. By 1981 alligators in Louisiana were downlisted to threatened, and by 1987 the entire U.S. population was taken off the list of endangered species. The reclassification of alligators from endangered to threatened allowed state wildlife managers to develop programs to deal with nuisance alligators and promote the economic utilization of alligators through farming, ranching, and direct harvests. Today, alligators are an important economic resource throughout the southeastern United States.

The recovery of the American alligator is a striking conservation success story, and one that demonstrates the resiliency of crocodilian populations to hunting pressure. Alligators responded to hunting by becoming increasingly wary and difficult

Figure 2.3. Basking alligators in Paynes Prairie, Florida, January 2007. The American alligator is once again common throughout most of its historic distribution.

to find. Despite the large number of hunters after their skins, breeding groups of alligators remained in many areas and became the foundation that allowed a rapid recovery once hunting stopped. The key factor here was the fact that American alligator habitat remained abundant throughout the entire species range. A similar situation is seen for a variety of other species including the black caiman and the Morelet's crocodile. Once widespread commercial hunting was controlled, populations entered a period of recovery. Species like the Chinese alligator, with little or no remaining habitat, represent a very different conservation challenge.

3

The History of Crocodilians in China

The alligator family (Alligatoridae) is a group of crocodilians that includes not only the two species of alligators but also the more tropical caiman, a New World assemblage of six species that ranges from southern Mexico to northern Argentina. While the caiman group is more speciose, the iconic species in this family, and the one best known from a biological point of view, is the American alligator (*A. mississippiensis*). The Chinese alligator is remarkable in that it is the only member of the family found outside the New World. In this chapter we review the history of the alligator in China, beginning with a summary of what is known about its evolutionary history and how the species may have made it to Asia. We also consider the past distribution of the species, particularly what is known or can be surmised about its range in the Holocene. We finish with an examination of some of the historical references to alligators in China, beginning with the ancient texts and ending with an account of the scientific description of the species in the nineteenth century and the scientific studies of the species made prior to 1990.

EVOLUTIONARY HISTORY OF CHINESE ALLIGATORS

The evolutionary origins of the Alligator family can be traced back to the Cretaceous (65 to 145 mya) of North America, at a time when the world was an ice-free hothouse and dinosaurs were the dominant megafauna. During the early Tertiary, a period of uneven but gradual global cooling, there appear to have been multiple dispersal events of alligatorids from North America to Eurasia (Brochu 2003), which led to a diversified group of species in what is today Europe. But these lineages eventually died out. Currently all members of the alligator family, with the exception of the Chinese alligator, are found in the New World. The caimans likely originated from early North American alligatorids that dispersed to South America, possibly as early as the late Cretaceous, and were in South America by the Paleocene when the two continents separated (Brochu 1999). The earliest known fossils for the genus *Alligator* come from the Oligocene (23 to 34 mya) in western North America and are referable to the species *Alligator prenasalis* (Brochu 1999). Three other North American fossil species are known from the Tertiary: *A. mcgrewi*, *A. olseni*, and *A. mefferdi* (Figure 3.1); the phylogenetic relationships among *A. sinensis*, *A. mississippiensis*, *A. olseni*, and *A. mefferdi* are, however, still incompletely resolved (Snyder 2007). Another fossil, *A. thompsoni* from Nebraska and Texas, may be a synonym of *A. mississippiensis*, and if so it provides fossil evidence of American alligators as far back as 14 million years (Malone 1979; Brochu 1999). There is a good fossil record of both *A. mississippiensis* and *A. sinensis* from the Pleistocene (Brochu 1999).

Given an evolutionary origin in North America, how alligators got to Asia has been a matter of some interest and debate. Most of the known fossil Eurasian alligatorids are from the early Tertiary and are not closely related to the genus *Alligator*, so the Chinese alligator must have originated from a separate dispersal from North America. This seemingly extraordinary disjunct distribution is not actually atypical among a number of groups of plants and animals, and has been of great interest to biogeographers for more than two centuries. It was first described in 1716 by Joseph Francis Lafitau, a French Jesuit who, after reading an account of the ginseng plants in Manchuria, found them growing near Montreal. The floristic affinities between eastern North America and eastern Asia were subsequently described by a number of botanists, and similar distributions have also been found in a variety of other groups ranging from fungi and arachnids to freshwater fish and salamanders (J. Wen 1999). Intercontinental species pairs of plants found in the eastern United States and Eastern China are common, and during our surveys for alligators in Anhui Province JT was pleasantly surprised that he could identify, at least to genus, many trees in Anhui Province that are also common in the eastern United States, including maples (*Acer* sp.), sycamores (*Platanus*), sweetgums (*Liquidambar*), magnolias (*Magnolia*), and pines (*Pinus*). A common wetland plant frequently associated with alligators, water lotus (*Nelumbo* sp.), is also shared between the two regions.

It is now generally accepted that the affinities between the flora and fauna of the eastern United States and eastern Asia are the result of what was, in the early Tertiary, a widespread temperate forest biome that spanned the two continents. Temperate species were able to disperse across the northern reaches of the world via both

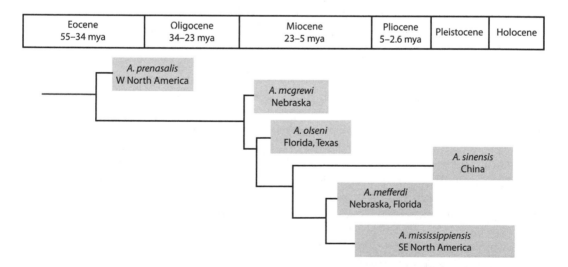

Figure 3.1. Proposed phylogenetic relationships of fossil and living alligators. Based on Brochu 1999; Snyder 2001.

North Atlantic and Bering land bridges during the climatically mild early Tertiary. The shortest and most likely of these two routes for alligators would have been across Beringia, as fossil alligatorids from the Tertiary (possibly in the genus *Alligator*) are known from Oregon and California (Brochu 2003). However, the general trend of our planet's temperature since the demise of the dinosaurs has been that of cooling, from the steamy world of the Cretaceous to the ice-house world of the Pleistocene (Markwick 1998). As global temperatures declined in the later Tertiary and Quaternary, these northerly dispersal routes were covered with boreal forest, isolating cold intolerant species in the warmer regions of the two continents. Periods of mountain building, cooling, and a dramatic drying of the climate in western North America in the later Tertiary and Quaternary resulted in intermediate forms disappearing, leaving similar species groups in the eastern United States and China. That alligators were able to follow the same general biogeographic pattern as many of these other species is not so unusual given that there is evidence from the late Cretaceous through the mid-Eocene (ca. 45 to 50 mya) of fossil crocodilians found up to at least latitude 60° N (Markwick 1998).

The fossil record in China is very incomplete, but it does provide us with important information. There have been a number of reports of early fossil remains of alligatorids in China, but the relationship of these to the *Alligator* lineage has not been adequately examined. In China, fossil alligatorids from the Paleocene uncovered in Nanxiong, Guangdong Province, have been assigned to *Eoalligator chunyii* (Q. Xu and Huang 1984). Unidentified alligatorid fossils from the late Eocene to the early Oligocene are widely distributed throughout China (B. Chen 1990a) including the far northwest (Junggar Basin in Xinjiang Uygur Autonomous Region) and the far south (Guanxi Zhuang Autonomous Region and Hainan Island). Perhaps the best-represented Asian fossil Alligatorid is *Alligator lucius*, from Shandong Province in deposits that are from the mid-Miocene (Li and Wang 1987). The remains consist of a nearly complete skull and parts of the postcranial skeleton, and the authors conclude that it is most closely related to *A. mcgrewi*. Brochu (pers. comm.) considers *A. lucius* to be a true alligator, and in fact to be within the clade of *A. mississippiensis* and *A. sinensis*. The Shanwang Basin deposits where *A. lucius* was found have been estimated to be between 15 and 18 million years old (J. Yang et al. 2007). Other uncorroborated mid-Miocene remains of alligatorids have been reported from northern Jiangsu Province (Chow and Wang 1964) and in the late Pliocene of Japan (B. Chen et al. 2003).

So when did alligators make it to Asia? Based on the fossils of *A. lucius* we know that there were true alligators in China at least by the mid-Miocene; how much earlier they got there is still unresolved. An early divergence between American and Chinese alligators would have allowed dispersal to Asia during the climatically mild Early Tertiary (Oligocene). The timing of the divergence between American and Chinese alligator lineages based on molecular data generally supports an early split between the two. Based on mitochondrial DNA sequence data, X. Wu et al. (2003) estimated the split occurred 50.9 mya, and Mo et al. (1991) estimated 45 mya using DNA–DNA renaturation kinetics data. Two other estimates of divergence time using mtDNA are approximately 76 mya (late Cretaceous) (Janke et al. 2005) and 47 to 53 mya (early Eocene) (Roos et al. 2007). These last two dates were produced by the same lab, and the differences between them are unexplained in Roos et al. (2007).

These estimates rely on the use of a "molecular clock" to time divergence based on the amount of accumulated genetic differences in the two samples, but calibration of molecular clocks can be difficult. These estimates for the timing of divergence in *Alligator* are questionable because they indicate the split occurred well before the earliest known date of *Alligator* fossils (ca. 23 mya). Brochu (2003) considers that combined fossil and molecular data indicate that A. *sinensis* and A. *mississippiensis* shared a common ancestor as recently as 20 mya. Q. Xu and Huang (1984) also proposed that the split between North American and Asian alligator lineages took place during a relatively brief period of warming in the Miocene. Based on the estimated dates of the Shangwang Basin containing fossils of A. *lucius*, alligators were in Asia at least by 15 mya, but how much earlier than this remains open to speculation. Alligators, the most cold tolerant of the crocodilians, could have survived in areas with cold winters by burrowing but would have required relatively mild summer temperatures to survive. Snyder (2001) evaluated the possibility that alligators may have entered eastern Asia during the period ranging from the mid-Oligocene to mid-Miocene but concluded that detailed analyses of paleoclimate in the region make most interpretations guesswork. Nevertheless, some information suggests that an Aleutian land bridge or island chain existed 1,500 km south of the Bering Strait during this period (15 to 30 mya). If so, it could have provided a dispersal route to Asia that is consistent with the most broadly accepted estimates of the evolutionary separation between A. *mississippiensis* and A. *sinensis* (ca. 20 mya).

PAST DISTRIBUTION

To reconstruct the historical distribution of alligators in China it is important to understand Pleistocene and Early Holocene climate shifts. The Pleistocene geological epoch began some 1.8 mya when the world was plunged into a series of ice ages. Global temperatures plummeted and the enormous growth of glaciers tied up vast quantities of water as ice, resulting in a dramatic drop in ocean sea levels and exposing large expanses of the continental shelf that today are underwater. Four major glacier cycles have been identified, by far the best known being the last one, referred to as the Wisconsin in North America and the Würm in Europe. Between the glacial peaks were warmer interglacial periods. The world's flora and fauna were dramatically impacted by these climatic shifts, which altered temperature and rainfall patterns globally. Reptiles, which as ectotherms are particularly sensitive to temperature changes, shifted their distributions in accordance with these climate events. At the peak of the last glacial cycle, conditions in Central China and the lower Yangtze River valley were very different from what they are today. Late Pleistocene paleoclimate reconstructions have found that during the peak of the last glaciation (ca. 20,000 years BP), North Central China was birch-shrub tundra that supported woolly mammoths and wooly rhinoceros, and Central China's coastal lowlands were dominated by cold tolerant coniferous trees, particularly spruces and firs (Winkler and Wang 1993). During this period the Yangtze River basin was too cold for alligators, which presumably were found farther south, probably in wetlands on the now submerged continental shelf of southern China. At that time the sea level along the eastern coast of China was about 120 m below its present level, and the shoreline was as much as 600 km farther east than at present (Winkler and Wang 1993). As a result, today's islands of Japan, Taiwan, and Hainan were part of the mainland. During this period, south China is thought to have retained a subtropical-warm-temperate climate, and it is likely that cold intolerant species assemblages, including the Chinese alligator, would have survived along river systems that drain what is today southern Zhejiang, Fujian, and Guangdong provinces.

Chinese alligator fossils have been uncovered from the middle Pleistocene in Anhui Province (Hexian) (Q. Xu and Huang 1984; B. Chen 1990a) from a site with a thermoluminescence date of about 190,000 years. These remains likely correspond to one of the interglacial periods when conditions in the lower Yangtze River valley were warm enough to support a warm-temperate fauna. Other species found at the same site include both northern (e.g., *Ursos arctos*) and southern (e.g., *Tapirus sinensis, Rhinoceros sinensis*) faunal elements (Bakken 1997). If alligators had moved south during periods of glacial advances, we would expect to find remains of Chinese alligators from more southern locations. One problem here is that at that time sea levels were much lower and many of the low elevation sites where alligators would be expected to be found are now underwater. Nevertheless, there is one report of finding alligator (*Alligator* cf. *sinensis*) remains in Guangdong Province, southern China (W. Huang et al. 1988). The bones were among a Middle to Late Pleistocene assemblage that included other aquatic reptiles (*Chinemys reevesii*) as well as mammals including pandas (*Ailuropoda melanoleuca*), pigmy elephants (*Stegodon orientalis*), bears (*Ursus thibetanus*), and tigers (*Pantera tigris*), which together represent a transitional sub-tropical/temperate fauna along the Xijiang River (ca. lat 22° N). Although the evidence is at best fragmentary, together these findings suggest that alligators, and other faunal assemblages, moved up and down along the eastern Asian coastal habitats repeatedly during the waxing and waning of the Pleistocene glacial periods.

The end of the global cold period known as the Younger Dryas approximately 11,500 years ago marked the transition from the Pleistocene to the warmer Holocene and ushered in a gradual shift of climate. During this period there was a rapid rise in sea level as the glaciers melted. Sea level was actually somewhat higher than at present in the Yangtze Delta region around 7,500 years ago. Human social organization also began a dramatic shift from nomadic groups of hunter-gatherers to the first agriculture-based villages. It is during this transitional period in human culture, generally referred to as the Neolithic, that we begin finding more archaeological evidence of humans, the remains of the animals they consumed, and their incipient agriculture. The warming trend, while uneven, continued, and in fact during this early Neolithic period, from 6,000 to 7,200 years BP, there is evidence that average air temperatures were up to 3°C warmer than they are today (Winkler and Wang 1993), allowing cold intolerant species to extend their range farther north along the Chinese coastal plain. Reconstructions of paleovegetation zones suggest that plant communities shifted 2° to 3° to the north during this period (Y. Shi et al. 1993). While the warming trend would have allowed alligators to extend their range, the limits of northern distribution for alligators may be determined to a greater degree by the lowest winter temperature, rather than mean annual temperature (Chapter 5). One of the unusual characteristics of the warming period for eastern Asia in the early Holocene was that winter minimum temperatures rose considerably more than mean annual temperatures and were some 6°C warmer than today (Y. Shi et al. 1993). The North China Plain was also a much rainier place 6,000 to 7,000 years BP, with some 450 mm of

additional rain creating large lakes on the plain (Y. Shi et al. 1993). The combination
of warmer year-round temperatures, milder winters, and increased rainfall allowed
a subtropical fauna, including Asian elephants (*E. maximus*) and Javan rhinoceros
(*R. sondaicus*), to thrive in Eastern China up to 41° N during this time (Y. Shi et al.
1993). There is archaeological evidence that alligators also moved north but apparently not as far north as some of the subtropical mammals.

In Central China the earliest known agricultural settlements date back to 8,000
to 9,000 years BP (Pengtoushan and Bashidang assemblages); over the following millennia they spread throughout much of the lower Yangtze River valley, based on rice
cultivation, and the Yellow River valley, with millet as the main crop (T. Lu 1999).
One of the best studied of these Neolithic cultures is at Hemedu, along the southern
shore of Hangzhou Bay in Zhejiang Province. Studies in the 1970s uncovered the
remains of several Neolithic settlements that date to approximately 7,000 years BP.
These were prosperous villages that subsisted on farming, rearing domestic animals,
and hunting and fishing. Excavations at Hemedu have uncovered a wealth of animal remains that include numerous subtropical species such as elephants, Sumatran
and Javan rhinoceros, pangolins, freshwater and sea turtles, and Chinese alligators
(T. Lu 1999). Alligator remains dated from 5,000 to 7,000 years BP have also been
found at another site, referred to as Ts'ao-hsieh-shan by K.-C. Chang (1987); it is
located on the southern borders of the Yangdeng Xihu lake, near the modern city
of Suzhou (Jiangsu Province) (K.-C. Chang 1987). Findings from Kuahuqiao, an
even older (7,000–8,000 years BP) Neolithic village site along the lower Qiantang
River in today's Zhejiang Province also include remains of alligators (L. Jiang and
Liu 2005).

Remains of an extensive subtropical and warm-temperate fauna at these and
other early Neolithic sites in the lower Yangtze River basin suggest that these species
expanded north into these regions fairly quickly after the end of the cool Pleistocene period. In fact contemporary evidence indicates that alligators were found even
farther north by this time. Alligator remains, including teeth and bony osteoderms
(L. Liu 2004), have been recovered from the Jiahu archaeological site in southern
Henan Province, which is considered to be among the earliest representations of the
early Neolithic Peiligang culture. The Jiahu settlement (lat 33°36' N) was established
as early as 7,000 BC and was based primarily on the cultivation of millet and rice, the
latter indicating that the local climate was warmer and moister than farther north
in the Yellow River valley (L. Liu 2004). Jiahu is more than 2° farther north than the
Yangtze basin sites and is located within the Huai River basin, the principal basin
located between the Yellow and the Yangtze rivers, at about 68 m elevation. These
alligator remains were found in association with a variety of other items related to
cooking or food storage, which strongly suggests that the alligators were killed locally
for food.

Even farther north than Jiahu, skeletal remains of at least 20 alligators, estimated
to be 1 to 1.5 m TL, have been found at the Wangyin site in the Huang-huai Plains
of Shandong Province (B. Zhou 1982) (lat 35°33' N, long 116°50' E), which dates to
about 6,000 years BP. These partially burned remains, which were found together
with those of fish and turtles indicating that they were caught and eaten (B. Zhou
1982), are the most northerly confirmed remains of Chinese alligators suggestive
of locally killed animals. B. Zhou (1982) cites this as evidence of Chinese alligators

being found in the Yellow River drainage, an area from which they subsequently disappeared as a result of hunting and the conversion of wetlands to agriculture. However, while this site is quite far north and close to the Yellow River (approximately 90 km to the northwest at its closest point), it is actually in the drainage of the more southerly Huai River. Although residents of this ancient village may have hunted and fished in the Yellow River, the finding of alligator remains at Wangyin cannot be used as evidence for their presence in the Yellow River basin.

Remains of alligators have been found even farther north at other Chinese Neolithic sites in present-day Shanxi and Shandong provinces in the Yellow River drainage. However, unlike evidence at the previous sites, these were man-made objects predominantly associated with human burials, including drums made using the skin of alligators. According to some studies, alligators may have held special significance for these ancient peoples (Chapter 4), and these remains are less useful for determining the past distribution of alligators because the alligator-skin drums could have been made from locally caught animals, or the skins could have originated from farther south and been traded. The existence of this type of trade has been noted in ancient records that indicate alligator skins were sent from Hangzhou as a tribute to the emperor (Fauvel 1879), and L. Liu (2004) suggests that the presence of the alligator-skin drums is evidence of long-distance trading of ritual items among elite members of these early Chinese societies.

The accumulated archaeological evidence shows that over the last 1,000 to 7,000 years, Chinese alligators were likely found in three major river systems in Eastern China, the Qiantang River and associated drainages that flow into Hangzhou Bay in today's Zhejiang Province; the Yangtze River; and the Huai River basin over a latitudinal range from 30° to 35° N. While archaeological remains from sites like Wangyin suggest that alligators were found up to the northern reaches of the Huai basin, only a relatively short distance from the southernmost sections of the Yellow River, we find no conclusive evidence that there were populations of alligators living in the lower Yellow River itself. Nonetheless there is some reason to believe they may have been found there during the warm Neolithic period. The Yellow River has the dubious honor of being the world's most silt-laden major river system. Levees built to contain it have reportedly failed more than 1,500 times and the river's course has changed more than 20 times in the last three millennia. In AD 1194 a catastrophic breaching of the levee and flooding to the south of the river forced water from the Yellow River to flow south into the Huai River basin. The mouth of the Yellow River shifted some 300 km south from the Bohai Sea into the Yellow Sea in northern Jiangsu Province. The river maintained this course for the next 700 years, reverting back to its previous configuration in the late nineteenth century. This also happened on at least two occasions in the Neolithic period (L. Liu 2004), which means repeated connections between the river systems could have facilitated the northward migration of alligators into the Yellow River.

The elimination of Chinese alligators from the northern portions of their former range is likely the result of the expanding human influence in the region, coupled with a gradual regional cooling trend. The earliest reported disappearance of alligators south of the Yellow River is at the southern extreme of the species' historical range, from the vicinity of Shaoxing, on Hangzhou Bay, in river systems to the south of the Yangtze. Subfossil remains (Neolithic–Recent) from the vicinity of Yuyao (B. Chen

1990a) indicate that alligators were found in central coastal Zhejiang Province. It would not be unreasonable to expect that some alligators populations may have remained in low-elevation wetlands in Zhejiang and perhaps northern Fujian provinces in the late Pleistocene while warming climates and rising sea levels enabled the species to move into the Yangtze and even farther north. Historical records suggest alligators were extirpated from the Shaoxing area around AD 1201 (B. Chen 1990a).

EARLY HISTORICAL ACCOUNTS

China is unique in terms of the extent and antiquity of its historical record, providing an exceptional opportunity to examine interactions between people and wildlife and how those interactions have changed over the millennia. Chinese historical accounts provide a rich trove of information on the past distribution of some of the country's more noteworthy fauna such as tigers and elephants (Elvin 2004). Unfortunately, the use of old written accounts to piece together the former distribution and abundance of alligators is somewhat more difficult. Old records of large mammals can usually be easily referenced to the species in question. But accounts of alligators may be lumped together with references to other crocodilians that were found in the more tropical south of the empire (such as the larger and more dangerous estuarine crocodile), to snakes, or even to mythical accounts of dragons or other water deities. Early descriptions of T'o, or earth dragons, that were more than 3 m long may refer to crocodiles or simply be an exaggeration of the size of alligators. It is not, however, inconceivable that Chinese alligators once grew larger than they do today (ca. 2.5 m; Chapter 5) and that their antagonistic cohabitation of the landscape with humans has exerted a strong evolutionary selective pressure toward small size and cryptic behavior.

Historical records also refer to crocodilian-like animals far to the north of the species' current distribution. Here, there can be no mistaking the species being referred to: it could be no other than the relatively cold tolerant alligator. One of the early references to alligators in Shandong Province comes from an account of the famous Chinese philosopher Confucius. The book of Chuang Tzu, attributed to the influential Taoist writer of the same name (ca. 369–286 BC) who was a native of the state of Meng, on the border of present-day Shandong and Henan provinces, noted that when Confucius was visiting the Lu Liang bridge, in Shandong, the water rose suddenly and flooded a great expanse of the surrounding area. The current was so swift that aquatic creatures, including the T'o (Chinese alligator), did not dare enter it to swim (Fauvel 1879).

The ancient distribution of the Chinese alligators reconstructed by H. Wen (1995) based on reports of crocodilian fossils and unspecified historical sources includes a number of sites from the Yellow River basin prior to AD 200 (Han Dynasty and earlier). Wen proposed that approximately 6,000 years ago Chinese alligators were found up to latitude 37° N and that this range had retreated southward to 35° N by 3,500 years ago, 33° N by 2,000 years BP, and 32° N by 1,000 years BP, or a reduction of approximately latitude 0.9° N for each 1,000 years. However, most of the more northerly locations cited by Wen appear to be archaeological sites where drums made of alligator skin have been found. Because it is likely that these valuable ritual items were traded or exchanged, sometimes over long distances (T. Liu 2004), they

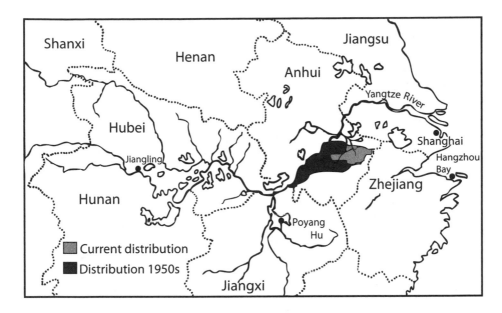

Figure 3.2. Approximate current range of wild Chinese alligators.

cannot be used to indicate the local presence of alligators. The northernmost distribution of Chinese alligators in historical times, however, remains imperfectly known and some of the references used by H. Wen (1995) suggest a more recent presence of alligators in the Yellow River basin, including a Song Dynasty (960–1279) mention of an alligator from Kaifeng (Henan Province). (Current distribution of Chinese alligators is shown in Figure 3.2.)

Oracle bone inscriptions are the oldest known form of Chinese writing, with some dating back to 4,800 years BP. These bones, usually the scapula of cattle or tortoise plastrons (Figure 3.3), were used for divination, the attempt to understand, and predict, events. The bones were placed on a fire until they developed cracks, the pattern of which was interpreted; the results and date of the divination were etched into the bone. The earliest references to crocodilians in China come from oracular bone inscriptions made during the Shang Dynasty (Yin portion, 1401 to 1154 BC; M. Zhang and Huang 1979).

Another source of early Chinese records are accounts that formed the basis for Confucianism and together are referred to as the Chinese classical texts. Dragons in China are first mentioned in the *Yijing* (or *I Ching*) and their habits as described in that work bear a striking resemblance to alligators (Chapter 4). Another of the classics, the Doctrine of the Mean, lists a number of aquatic creatures including one known as the *T'o*. The identity of *T'o* is hard to pin down. The earliest dictionary of Chinese terms was compiled by Robert Morrison, the first Protestant missionary in China, who had the Bible translated into Chinese by 1815. Morrison also compiled a Chinese-English dictionary, which translated *T'o* as "a large sea (?) animal, upward of 10 feet long, a species of fish" (Fauvel 1879). A subsequent English translation of the Chinese classics by James Legge, a noted Scottish sinologist and the first professor of Chinese at Oxford University, refers to *T'o* as an iguana. In 1879 Fauvel

Figure 3.3. A Chinese turtle plastron with inscriptions that was used as an "oracle bone" for divinations. It is from a 1996 series of Chinese stamps issued to commemorate the ancient capital of Anyang.

Figure 3.4. Chinese character (from Fauvel 1879) for *T'o*, the earliest form of which is said by Li Shih-chen, author of *Pen Tsao Kang Mu* (*The Great Herbal*), to have been derived from a pictogram representing the head, body, feet, and tail of an alligator.

argued convincingly, based on ancient descriptions of the mysterious creature, that the *T'o* was actually a Chinese alligator, an animal unknown to Morrison and Legge (Figure 3.4).

Brief mentions of the *T'o* are sprinkled throughout other Chinese classics. In the *Shijing*, or Book of Poetry, there is a reference to how the skin of *T'o* was used for enormous drums. The *Liji*, or Classic of Rites, was a bit more specific, noting that in the ninth moon people would capture *T'o* and send them to the imperial palace where drums were made from their skins. Later commentaries on the classics note that the sounds made by drums are similar to the calls of the *T'o*, whose name is an onomatopoeic reference to its vocalizations (Fauvel 1879). Some of the oldest references to alligators may actually mention early attempts to breed them. In the *Zuozhuan* (Comments by Zuo), a commentary on Confucius's *Spring and Autumn Annals*, it was said that during the time of the Yellow Emperor (Huangdi) and Shun, two legendary leaders purported to live between 2600 and 2200 BC, some individuals had great skill in breeding "dragons," possibly a reference to *tu long*: earth dragon, the Chinese alligator (Laidler and Laidler 1996).

But the Chinese alligator was not the only crocodilian native to China in historical times, and some of the ancient references to *T'o* may be to other species. K. Zhao et al. (1986) suggest that more than 1,000 years ago the Malayan gharial (*Tomistoma schlegelii*) was probably widespread in southern China (Guangdong and Guangxi provinces). Fossils of two earlier species of *Tomistoma* (*T. petrolica* and *T. taiwanicus*) have been found in Guangdong and Taiwan, and two skulls of the modern-day species *T. schlegelii* were found (in 1963 and 1973) in Shunde County, Guangdong. J. Fu (1994) and H. Wen (1980) suggest that accounts of crocodilians in southern China in the Han Dynasty (206 BC–AD 220) may refer to either the estuarine crocodile (*C. porosus*) or the Malayan gharial (*T. schlegelii*), which were at that time found as far north as Fuzhou (Wan Lin River). Historical references to crocodiles in Guangdong were common from the second to the fifth century AD and became well known to people throughout China from the writings of Han Yu during the Tang Dynasty (618–907) (K. Zhao et al. 1986). *Tomistoma* was probably extirpated from China in the Yuan (1260–1368) or Ming dynasties (1368–1644)

(K. Zhao et al. 1986), but *C. porosus* likely survived in parts of southern China until more recent times.

In his remarkable description of the Chinese alligator, Fauvel (1879) was able to piece together accounts of crocodilians in early China by referring to Chinese dictionaries and other reference materials. Some of these descriptions may refer to alligators, some to crocodiles, and some to a mixture of the two. Early references to an animal called *Ngo* in southern China or parts of what is now Southeast Asia undoubtedly concern crocodiles. During the Liang Dynasty (502–556), in the kingdom of Lin Yi (present-day Vietnam) moats around the capital were said to be stocked with *Ngo*. A contemporary method of judging presumptive criminals was to throw them in the water with the *Ngo*; if they survived three days they were declared innocent. An allegorical account of how the wild regions of southern China were tamed concerns the renowned poet and statesman Han Yu. During the Tang Dynasty (768–924) he was sent to serve as governor of Chao Chou, a semibarbaric region in what is today Guangdong Province. Upon his arrival he was told by the local people about a small lake or marsh that was filled with *Ngo* that hatched from eggs and grew to 20 feet long. The *Ngo* were a menace because they killed and consumed cattle and other domestic animals. A few days later Han Yu visited the location and, after getting the attention of the *Ngo* by tossing a pig and a sheep in the water, he addressed the denizens of the pond:

> Under former rulers you have been allowed to remain here, but under the reign of our virtuous emperor you cannot be tolerated and you must leave his empire. How could you be permitted to live here in peace when you are molesting the people [by] fattening on their domestic animals and increasing daily in number? I have come to rule this country in the name of the sovereign and as I am myself much afraid of you we must part company. At the south of this place is an immense sea in which fishes as large as whales as well as those as small as shrimps and sprats can live in peace. You can easily go there in a day but I give you from three to seven days to go. If after that period you are still found here I shall be compelled to bring with me some good archers with strong bows and poisoned arrows and declare against you a merciless war.

That afternoon a great storm arrived. After several days the wind had blown all the water from the lake. Afterward it remained dry and the *Ngo* were never seen there again.

Another of the references used by Fauvel was the great Chinese medical work *Pen Tsao Kang Mu* (*The Great Herbal*), compiled in the Ming Dynasty by Li Shih-chen (1518–1593), who is considered the father of China's herbal medicine. Here *T'o*, said to belong to the family of dragons and to be 10 feet long, is illustrated (Figure 3.5). The author notes that the *T'o* will dig deep burrows, where they can be captured using baited hooks, and that people in "the south" are very fond of the flesh which they serve as a delicacy at every wedding feast. Shih-chen Li quotes earlier works, including one by Chen Chi, that say the *T'o* is "of a sleepy nature, often closing its eyes" but that it is also "gifted with great strength and can burrow in the banks of rivers." In what may be accounts of estuarine crocodiles in the south of China, it was also reported that the animals grew 10 to 20 feet long, had backs and tails covered with armor, cried at night, and were feared by mariners.

Figure 3.5 The *T'o*, as illustrated in the sixteenth-century Chinese medical work *Pen Tsao Kang Mu* (*The Great Herbal*). From Fauvel 1879.

EUROPEANS ARRIVE IN CHINA

The Venetian trader and world traveler Marco Polo is generally recognized as the first European to have compiled a written record of China, which is arguably the world's most famous travelogue. Polo, traveling east along the Silk Road, visited China during the late thirteenth century, at the very end of the Song Dynasty, widely considered to be the greatest Chinese period in terms of technology, industry, and the arts. Polo, a gifted linguist, became a favorite of Kublai Khan, in whose court he served for 17 years. There has been considerable scholarly debate about the authenticity of Polo's accounts (they were dictated largely from memory in 1298 while he was in a Genoese prison three years after his return to Italy). Whatever the case, the resulting book, *The Description of the World; or, The Travels of Marco Polo*, became a medieval best seller with its tales of the exotic Mongol Empire and strange lands in India and Africa. Polo's accounts of the great wealth and splendor of Cathay (China) include a visit to Hangzhou, just south of the mouth of the Yangtze and at that time the capital of the southern Song Dynasty. Prior to the Mongol invasion in 1279, Hangzhou was the world's largest city, with a population that may have been in excess of 2 million. Polo was amazed at the number of boats on the Yangtze, declaring that "this river goes so far and through so many regions and there are so many cities on its banks that, to tell the truth, in the total volume and value of the traffic on it, it exceeds all the rivers of the Christians put together plus their seas."

Polo also made reference to some of the region's unusual fauna. It is clear that he saw some large reptiles in China.

> In this province are found snakes and great serpents of such a vast size as to strike fear into those who see them, and so hideous that the very account of them must excite the wonder of those who hear it. I will tell you how long and big they are.
>
> You may be assured that some of them are ten paces in length, some are more and some less. And in bulk they are equal to a great cask, for the bigger ones are about ten palms in girth. They have two forelegs near the head, but for foot nothing but a claw like the claw of a hawk or that of a lion. The head is very big, and the eyes are bigger than a great loaf of bread. The mouth is large enough to swallow a whole man, and is garnished with great (pointed) teeth. And in short they are so fierce looking and so hideously ugly, that every man and beast must stand in fear and trembling of them. There are also smaller ones such as of eight paces long, and of five, and of one pace only. The way in which they are caught is this: you must know that by day they live underground because of the great heat, at night they go out to feed and devour every animal they can catch. They go also to drink at the rivers and lakes and springs. And their weight is so great that when they travel in search of food or drink, as they do by night, the tail makes a great furrow in the soil as if a full ton of liquor had been dragged along. Now the huntsman then takes them by a certain gyn which they set in the track over which the serpent has passed knowing that the beast will come back the same way. The plant a stake deep in the ground and fix on the head of this a sharp blade of steel made like a razor or lance point, and then they cover the whole with sand so that the serpent cannot see it. Indeed the huntsman plants several such stakes and blades on the track. On coming to the spot the beast strikes against the iron with such force that it enters his breast and rives him up to the naval, so that he dies on the spot (and the crows seeing the brute dead begin to caw, and then the huntsman knows that the serpent is dead and comes in search of him).
>
> This then is the way these beasts are taken. Those who take them proceed to extract the gall from the inside, and [he] sells [that] at a great price; for you must know it furnishes the material for most precious medicine. Thus if a person is bitten by a mad dog, they give them but a small pennyweight of this medicine to drink, he is cured in a moment. Again if a woman is hard in labor they give her just such another dose and she is delivered at once. Yet again if one has any disease like the itch, or it may be worse, and applies a small quantity of this gall he shall easily be cured. So you see why it sells at such a high price. They also sell the flesh of this serpent, for it is excellent eating, and the people are very fond of it. And when these serpents are very hungry, sometimes it will seek out the [lairs] of lions or bears or other large wild beasts and devour their cubs, without the sire and dam being able to prevent it. Indeed if they catch the big ones themselves they devour them too; they can make no resistance.

Although Marco Polo refers to the animals as serpents, they are more likely crocodilians or a jumbled mixture of remembered animals that included crocodilians. These accounts of "serpents" were from Polo's visits to the province of Carajan, which today is Yunnan Province in southern China. In this region the native crocodilians were likely to have been estuarine crocodiles (*C. porosus*) or Malayan gharials (*Tomistoma* sp.). But his reference to the animals spending the day in underground burrows suggests he may have seen, or heard stories about, the Chinese alligator, a prolific burrower.

Not until 1653 did another European describe Chinese crocodilians, when the priest M. Martini did so in his work *Novus Atlas Sinensis*, published in Vienna. In 1640 Martini set out from Rome; he arrived in China in 1643 and worked there as a missionary until 1650. During his long return voyage he collated his valuable historical and cartographical data on China. Martini describes a city in Quangsi (= Guangxi) where the king kept 10 crocodiles in a small lake. Accused criminals were thrown to the crocodiles, and it was said that the innocent were left untouched while the guilty were devoured. As this was another southerly location in China and the crocodilians were apparently quite large and capable of eating people, the species in question is likely the estuarine crocodile. However, Martini also refers to animals in the more northern province of Chingkiang (= Jiangsu), within the historical range of Chinese alligators, where a river was inhabited by crocodilians that were considered to be a great plague to the people.

DESCRIPTIONS FOR SCIENCE

Throughout the Ming and Qing (1644–1911) dynasties, China remained closed to trade with European countries. Not until 1842 did the Treaty of Nanjing open China to the outside world. That treaty came at the end of the first of the so-called Opium Wars, which were fueled by the British desire to sell opium from its colony in India to the Chinese, in exchange for tea. Five ports, including Shanghai, were opened to British residence and trading. This was followed by similar agreements with France, Germany, and the United States. After the Second Opium War, an 1858 treaty opened another 10 ports to European trade and gave Western ships the right to freely navigate on the Yangtze River. As a result, in the mid-nineteenth century, there was an influx of Europeans into Eastern China. These were mostly merchants and others in pursuit of trade, but Christian missionaries and diplomats also arrived, some of whom were keenly interested in the natural history of the region. Among these was Père Armand David, a French Catholic missionary who was the first Westerner to describe the giant panda, and Pierre-Marie Heude, another Roman Catholic priest from France, who founded a natural history museum and devoted much of his time to the study of softshell turtles (E. Zhao and Adler 1993). But it was a third Frenchman, Albert-Auguste Fauvel, a customs officer and amateur naturalist stationed in Shanghai, who first brought the world's attention to the Chinese alligator.

Even before Fauvel's arrival on the scene there had been some tantalizing hints of a crocodilian living in the Yangtze. And while Fauvel made an exhaustive effort to summarize previous accounts of alligators in the ancient Chinese texts, the earliest contemporary account of Chinese alligators by a westerner, made by a Mr. E. A. Reynolds nearly escaped his attention. The story came to Fauvel's notice while his monograph was being published, with Reynolds writing a short account of his story to Fauvel, who then was able to stop the presses and include the interesting tale in his publication.

Reynolds was a well-to-do trader living in Shanghai, and in April 1853 he lent his "house-boat" to T. T. Meadows, a British consular officer. Meadows was accompanying the British governor of Hong Kong, S. G. Bonham, on a diplomatic mission to Nanjing to meet with the leaders of the Taiping Rebellion. Although little known in the West, the Taiping Rebellion was the world's bloodiest civil war (an estimated 20 to 30 million died), and among all armed conflicts it is surpassed only by the Second World War in terms of human carnage. The "rebellion" was led by Xiuquan Hong, a self-proclaimed Christian messiah who declared that he was a brother of Jesus Christ. Hong's ambition was to establish a Kingdom of Heavenly Peace and to do this he and his followers battled the Chinese imperial Qing Dynasty across southern and central China.

In March 1853 the rebels had taken the city of Nanjing and made it their capital. As Reynolds, Meadows, and Bonham were traveling on the Yangtze toward Nanjing, they passed near Silver Island, next to the modern city of Zhenjiang. Believing they saw bodies floating in the river, they stopped at the island. What they thought were corpses were instead wooden Buddhist carvings that had been thrown into the river by the rebels. Although the island's temples and other structures were destroyed, they found a Buddhist priest amid the ruins. Reynolds wrote:

> He took me to a pond or a small lake, taking a small bowl of rice and a switch with him, with the latter he beat the water crying "ado, ado," presently an Alligator or Crocodile came towards where we were standing, and while still in the water opened his mouth into which the priest threw the contents of the bowl: the Alligator backing himself into the water again. I was quite unprepared for such a sight, and was a little alarmed at first. I should say the animal was of a good size, but as his open mouth only came out of the water, I could not see how long it really was.
>
> Returning to the temple, I noticed a very fine etching, of an Alligator, also a long inscription cut in the slate tablet, I dug this out intending to bring it away, but on getting it to the ground it was found it would require several men to carry it off, so I had to abandon it. I, however, think it would have been replaced by the priests, and may be now in the wall, from whence I removed it. I should think one could get a full description of this Alligator from the inscription.
>
> A friend of mine, Captain Elsworthy, who was in the Taou-tai's fleet blockading the river, told me he had frequently been shewn this same Alligator by the priest some months after; and they told him it went away in the winter (more likely only buried itself deep in the mud at the bottom of the pond), and returned in the spring and remained on the Island during the hot months.

Later, Reynolds wrote: "Three years ago (1876), while ascending the Yangtsze in my steamer, being some miles below Nanking, my people were alarmed at a strange to them looking fish, which was close to the shore not more than ten feet from the steamer. I immediately ran to the side, and then saw an Alligator about eight feet long floundering in the wash caused by the paddle-wheels. I stopped the steamer with the view of capturing it, but it had disappeared. I am led to believe this same Alligator was since seen at Chinkiang, no doubt having drifted down by the current."

Another account by Europeans about Chinese alligators, and the first one that was actually published, albeit in a newspaper, was by a Mr. Goodwin. It appeared on 17 March 1869 in the *Shanghai Evening Courier* under the title "Crocodiles in China." Part of this entertaining account is presented here:

Figure 3.6. Etching of the alligator found on Silver Island in 1853 by Mr. E. A. Reynolds. From Fauvel 1879.

A little time ago, . . . we were interested in a report that a real true dragon had been imported and was to be seen by the curious in the Shanghai tea gardens. Naturally reports magnified the appearance and attributes of this extraordinary creature; nothing of the kind had been seen before; it had come out of a cave in the wild Kiangse mountains; could devour [a] child without distressing its thorax; and was eminently calculated to perform the supreme act of Chinese patriotism: exterminate the barbarian. A goodly crowd of foreigners went to see the monster. Armed with ten cent pieces . . . we found ourselves amidst the wonders of that most curious ground known as the tea gardens. Passing through the festive crowd who were spending their loose cash with mountebanks, peep shows, story tellers, and sweet meat vendors, we came upon a large space surrounded by a strong net in the middle of which was a canvas screen about two feet in length, and behind this was concealed the dragon. A notice at the entrance informed the public that he weighed so many catties and was a real horrible mountain dragon. Above the canvas screen could be seen the manly form of a coolie armed with a bamboo, who every five minutes appeared to be engaged in a kind of Poontung outrage (fight) with the monster within, and every five minutes by the gesticulations and cries of this gentleman it would appear that the dragon got the better of his antagonist, who disappeared defeated behind the canvas for a time before recommencing the performance. Truly there must be a "pucka" (proper) demon inside! Feeing the attendant, in we went, with a fore knowledge that we were going to be "done" and "done" we were. In an ordinary washing tub about three feet long was to be seen a poor miserable half dead common crocodile, or alligator as learned discriminators of species declared, who resembled his congener of the Nile about as much as a monkey does a man.

Later, the author goes on to say: "The party assembled, among whom were the learned both in law, language, medicine and science felt that they had come a long way to see nothing, but to console themselves, agreed that this wretched little crocodile having 32 teeth must be [a] new species, and that consequently both time and coin were well spent."

Goodwin's account, while interesting, did little to further the scientific understanding of Chinese crocodilians. However, the following year saw a note in a scientific journal alluding to the presence of a native crocodilian in the Yangtze region. This was published in 1870 by Robert Swinhoe, a British diplomat, in the 1870 *Proceedings of the Zoological Society of London* in an article titled "Notes on Reptilea and Batracians Collected in Various Parts of China." Swinhoe was the British consul in

Formosa (Taiwan), but traveled widely on the mainland, including long trips up the Yangtze to determine its navigability for steamships. Swinhoe was mainly interested in birds but he collected many specimens of amphibians and reptiles in Eastern China, describing many new species. Swinhoe spotted an animal on display in Shanghai in February 1869, observing what he called a young crocodile about four feet long. The exhibitors were apparently making so much money showing this animal, which they called a dragon, that they refused to sell it to him, and so its identity remained uncertain. It seems likely that this "dragon" was the same alligator mentioned in the colorful account by Goodwin.

Not until almost 10 years after Swinhoe's note did Albert-Auguste Fauvel turn his attention to crocodilians and describe the Chinese alligator in what is one of the classics of nineteenth-century herpetology (Figure 3.7). At the time, Fauvel was stationed in Shanghai as an employee of the Imperial Chinese Maritime Customs. Born in Cherbourg in 1851, he was the son of a well-known naval officer, Auguste Fauvel. Perhaps trying to follow in his father's footsteps, Fauvel joined the French navy but soon left due to poor eyesight. Because he spoke Chinese he was able to gain employment as a French customs officer in China. He arrived in China in 1872, first working in Beijing. He was later transferred to Yantai (then called Chefoo), a treaty port in Shandong Province, where he developed an interest in China's natural history (Bretschneider 1898). In 1877 he moved again, this time to Shanghai, where his naturalist predilections led him into an honorary curatorship at the Natural History Museum of Shanghai, a part of the North China Branch of the Royal Asiatic Society (Fan 2004).

While living in that sprawling hub of trade he began to hear rumors about "crocodiles" being sighted in the Yangtze River and set out to track down the origin of these stories. One of the first reports, in 1875, was of an alligator that had been caught in the Yangtze very close to the present city of Nanjing. The alligator was apparently kept a few days until a group of Chinese purchased it for fifteen dollars and gave it to some priests living on a nearby island (likely the same island mentioned in the account by Reynolds). Then in April 1878 a torpid alligator was brought to Fauvel by Lloyd E. Palm, the acting deputy commissioner of Chinese customs in Wuhu. The animal had apparently been dug out of its burrow from somewhere in the vicinity of Wuhu, purchased by Mr. Palm, and then taken to Shanghai. Upon receiving this animal, Fauvel, who had no suitable books with him in the museum, delved into the reference material belonging to a group of Jesuit priests in Shanghai. After carefully comparing it with the plates of known species of crocodilians he came to the exciting conclusion that not only was it was a new species, but one that was most closely related to the American alligator (which Fauvel referred to by its old name, *Alligator lucius*). Fauvel published a short note of the discovery in the *North China Daily News* on the ninth of May and included a request for more specimens:

> The museum has also received two new and important specimens: —a beautiful albatross (Diomedea Derogata?) shot by Capt. A. Croad near the Chusan group: and a living alligator (Alligator Lucius?) sent by Mr. J. L. E. Palm from Wuhu, where it was captured in the hills. Père Heudes having also seen Alligators near Ning-ko-fu, there is no more doubt about the existence of this saurian in the Yangtze waters. It differs considerably from the two specimens described in Cuvier and Duméril, and will likely

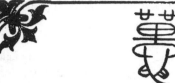

ALLIGATORS IN CHINA:

THEIR HISTORY, DESCRIPTION AND IDENTIFICATION.

READ BEFORE THE NORTH-CHINA BRANCH OF THE ROYAL ASIATIC SOCIETY, ON 13TH DECEMBER, 1878.

鼉　魚　鼍
鼓　鼈　鼉
逢　生　蛟
逢。　焉。　龍

The largest tortoises, alligators, crocodiles, fishes and turtles are produced in the waters.

The alligatorskin drums are resounding.

BY

A. A. FAUVEL,

IMPERIAL CHINESE MARITIME CUSTOMS,

Bachelier ès Sciences de l' Université de Paris and Honorary Curator of the Shanghai Museum.

Figure 3.7 The cover of Fauvel's treatise on Chinese alligators published in 1879.

prove a new variety. If any friend of science can succeed in sending us another live specimen, it will be forwarded to Paris for examination.

    —A. A. Fauvel, Hony. Curator

The advertisement worked and by that fall, Fauvel had four good specimens to use for the description of the new species, which was read before the North China Branch of the Royal Asiatic Society in Shanghai in December 1878 and published the following year.

Fauvel's first detailed account of the Chinese alligator was a privately printed monograph (Figure 3.7), probably published early in 1879. Fauvel had read his description of the new species before the Shanghai branch of the Royal Asiatic Society on 13 December 1878 and planned to publish it in the society's annual journal. However, he obtained permission from the society to print it privately before the journal was published. Later that year a slightly expanded version of the monograph came out in the *Journal of the North China Branch of the Royal Asiatic Society* (n.s., vol. 13) with one additional plate and additional information on Chinese and foreign historical references and natural history.

Fauvel's publications are an outstanding compilation of what was known about the alligator up to that time. He not only spent a considerable amount of time and effort collecting as many specimens as possible and reporting what was known about their natural history, he also delved into ancient Chinese texts looking for references to crocodilians and for the cultural significance of this animal. He found there was considerable confusion as to the nature of the old Chinese character for the alligator: 鼉 (*tuo*, or *T'o* as Fauvel writes it) and Fauvel began his monograph with a philological investigation of the meaning of this character and the biological identity of the creature to which it referred (Fan 2004).

The description of a new species of alligator was noted with considerable interest by scientists around the world. The only other alligator known to science was from the United States, the other side of the world, and all the related species in the family Alligatoridae were restricted to the New World. The finding of a true alligator in China was, according to Thomas Barbour (1910) at Harvard University, "one of the most interesting and remarkable facts which has ever been brought to the attention of zoogeographers." During the following decades a few more specimens were captured, and Chinese alligators began slowly making their way out of China and into European museums and zoos. Fauvel sent at least one stuffed specimen to the Paris Museum (Figure 3.8); it became the type specimen. In 1888 two live animals, purportedly from a pond near the Tienmu Mountains in Zhejiang, were purchased in Shanghai by the German consul O. F von Möllendorf and sent as a gift to Prince Bismarck. These were kept for a number of years at the Berlin Aquarium (Anonymous 1889). The same year two stuffed alligators collected at Kiu Kiang (= Jiujiang Shi, Jiangxi Province) by a Mr. Styan were sent to England; one specimen was presented to the British Museum. The following year, two live animals were received at the zoo in Frankfort and two more live alligators were sent to the Zoological Society of London and kept in their "menagerie." These latter animals were used by G. A. Boulenger (1890), curator at the British Museum, in a published description of additional features of the new species.

Figure 3.8 One of the authors (JT) holding the type specimen of the Chinese alligator in the Museum National d'Histoire Naturelle in Paris.

American museums were intensely interested in obtaining specimens of this new alligator. The first Chinese alligator reached the United States in 1903. The mounted specimen, captured somewhere near the mouth of the Yangtze, was a gift to Dr. W. T. Hornaday, then head of the New York Zoological Society (now the Wildlife Conservation Society), from a Capt. T. Golding. In 1910 Hornaday donated the alligator to the Museum of Comparative Zoology (MCZ) in Cambridge. Thomas Barbour, the MCZ's curator, wrote two short accounts based on his observations of specimens in the museum and others that came to the American Museum of Natural History in New York (Barbour 1910, 1922). Barbour's comments focused on the morphology and coloration of the alligator, although he also included several anecdotes based on the notes of Captain Golding. The alligators were referred to locally as *Yow lung* and *Tou lung*, which Golding indicated referred to types of dragons. Owing to persecution the alligators were rarely seen during the day, only emerging at night to hunt chickens or small dogs.

The first scientist to visit China in search of alligators was Clifford H. Pope, who was part of the American Museum of Natural History's Wulsin expedition (Figure 3.9). Pope was a herpetologist, and one of the biggest prizes to be collected was the Chinese alligator which Pope knew had been found principally from the vicinity of Wuhu, on the Yangtze River. When he first visited the area in the late summer of 1921, the area was flooded, and "the local fishermen had assured [him] that during flood times it is extremely difficult, if not impossible, to secure specimens" (Pope

1935). He returned to the area in March 1922, when river levels were much lower and collecting conditions more favorable. Many years later Pope (1955) gave a brief account of his trip:

> I was fortunate in being the first westerner to visit the home of this reptile and to secure a series of individuals for scientific study. . . . I was sent to Wuhu, a city a short distance up the Yangtze River from Nanking. It was from the region of Wuhu that most of the alligators appeared to come. Being strongly advised that the presence of a foreigner would make prices soar, I remained in the city and on March 14 sent out two scouts. Late on the nineteenth one of the scouts proudly strode in ahead of a rickshaw loaded with five securely bound alligators. Their sluggishness showed that they had been aroused from hibernation; the most active made feeble movements and roared a little.
>
> Our series of nineteen specimens was soon complete and I decided to visit the dens. Less than half a mile from a large village, the specimens had been dug out of burrows, each a foot in diameter and about five deep.

Some of Pope's notes from this trip were included in a subsequent publication by K. P. Schmidt (1927) and in a general book on his time in China (Pope 1940) where he reported that during a week's digging by local residents they had purchased the 19 alligators for a total of $19.60. The smallest was two feet long; the longest, four to

Figure 3.9 Clifford Pope in China with two of the alligators he collected for the American Museum of Natural History. From Pope 1940.

five feet long. The numerous burrows were said to be one foot in diameter at their opening and to terminate approximately five feet below the surface of the ground. (A "wild cat" was reported to have been found in one of the burrows.)

The burrows where the hibernating alligators were found were situated near the banks of the Chingshui River, approximately 11 km from the city of Wuhu, in an open, grassy plain through which the river meandered. Local residents reported that alligators were usually found together in small colonies. Besides the one from which they were collecting, only two other colonies were in the area (Schmidt 1927). Pope surmised that this plain was not cultivated because the area flooded deeply each year. Even at that time, Pope marveled at the ability of the alligator to survive in such a densely settled landscape.

As part of another zoological collection team, Sidney Hsiao (Xiao Zhidi), from the biology department of Hua Chung University in Wuchang, with support from the Rockefeller Foundation in China, was able to make more detailed observations on the behavior of the Chinese alligator. Like virtually all of his predecessors, Hsiao made his trip during the winter because this seasonally flooded area was more accessible during the low water period (Figure 3.10). Hsiao followed the field notes of Pope (in Schmidt 1927) to locate the area near Wuhu, where on a sunny December day, he traveled by rickshaw, with a cook from an American missionary school, to the nearby village of Ching Shui Ho. Using a local Seventh Day Adventist church as his base of operations, Hsiao spent his Christmas holidays looking for alligators. Offers to pay for alligators were circulated and soon a number of unfortunate animals were being excavated from their burrows by "many an energetic farmer." Hsiao provides a vivid description of the area:

> Traveling along the bank of Ching Shui Ho, or "Clear Water River," we observed that "the river flows from the south through Wuhu to the Yangtze," after winding its way among the low land covered with weeds. "At this time of the year the river flows between steep banks, perhaps twenty five feet high." there are many holes scattered along the banks of the river at different levels and one is tempted to think that any one of them might be the entrance into the home of an alligator, and thus worth exploring. But upon using one of the most common methods employed in collecting—asking the observant natives of the place for information—it was found that they were not inhabited by alligators. We reached the village town of Ching Shui Ho at dusk and stopped in the Seventh Day Adventist Church, to the pastor of which we had an introduction. In the evening word was passed around in the village that alligators were wanted. Early the next morning we went out with a number of children as our guides to the field for alligators. There are many ponds and creeks dried up at this time of the year. In one pond we located with help of the people who "possessed" the waste field and pond, the hibernating place of an alligator. The entrance to the hiding place of this alligator was on the northern side of the pond about two feet in diameter with its upper edge more than five feet from the level of the ground and its lower edge almost at the same level as the bottom of the pond. The entrance led into a bent tunnel about fifteen feet long. The male members of the family who claimed the pond, were engaged to make a vertical cut from the surface of the ground down to the tunnel. This was not the only place where alligators were found that day; for many an energetic farmer, after being notified of our desire to buy alligators, was also

digging for this animal. Before the day was over four large-size alligators had been procured. Beside the physical labor of digging through the dry soil for the hiding places, there was little difficulty in collecting the alligators. We had less fight against the reptiles than against the erroneous idea among the villagers that we had unlimited wealth for buying the alligators. One evening we were almost mobbed by the worse elements among the villagers. On the other hand the ferocity of reptiles did not exceed a sudden "roaring" and a showing of teeth in a few cases. As the Christmas holidays expired, we hired a local man as our collector. He continued to collect for us and secured nine mature and seven immature alligators by the end of January. The collector so firmly believed in the popular idea that "alligators crack turtles for food like beans" that he bought a basket of the shelled reptiles at five cents apiece and brought them to us with the alligators.

By the end of January they had 28 alligators packed into wooden boxes and shipped back to Wuchang, where a pond with "underground tunnels" was built so Hsiao could make observations on behavior and feeding habits. One of his interests was to determine whether the alligators would eat domestic pets, and he conducted numerous trials without success. He writes: "Furthermore, young puppies have been placed in the tunnels of the alligators for months without being molested in any way, but they seem to get along as friends with the alligators." During these same trials he found that the alligators would readily devour frogs, fish, and ducklings. Hsiao (1934) found no evidence of the alligators making a nest (as he knew American alligators

Figure 3.10 Historic photo of winter marshes near Wuhu. These marshes were the site of alligator dens from which the majority of Chinese alligators were collected in the late nineteenth and early twentieth centuries. From Pope 1940.

did); instead, he suggested that the female left her eggs "among the weeds in the waste fields and [leaves] them to hatch under the heat of the sun."

No further scientific work on alligators was done in China during the rest of the turbulent 1930s and 1940s. Following the 1949 Chinese Revolution and the creation of the modern Communist state, research on wild alligators in China began anew with the work of Cheng-Kuan Chu (Chengguan Zhu) and his student Chu-chien Huang (Zhujiang Huang), who carried out studies from 1951 to 1956, much of which is summarized in Chu (1957). The studies were again conducted on animals in the area adjacent to Wuhu and provided basic information on the ecology of alligators, particularly the structure of burrows.

Studies of Chinese alligators came to a halt during the Cultural Revolution (1966–1976) but were reinitiated soon thereafter by a group of Chinese scientists including Chu-Chien Huang, from the Institute of Zoology in the Academica Sinica in Beijing, and Bihui Chen, from Anhui Teacher's College in Wuhu. In 1979, Huang visited the United States and provided evidence that Chinese alligators remained in the wild, which led to a joint Sino-American study in 1981 with Huang, Chen, and Myrna Watanabe, an American scientist who had completed her PhD on the behavior of American alligators. By the 1980s most of the work in China focused on the captive breeding programs in Anhui and, to a lesser extent, in Zhejiang Province. Periodic surveys were carried out by the Anhui Forestry Bureau. These and other more recent biological investigations are discussed in Chapter 5.

龘

Alligators as Dragons

Among all kindes of Serpentes, there is none comparable to the
Dragon, or that affordeth and yeeldeth so much plentiful matter
in Historie for the ample discovery of the nature thereof.

—Edward Topsell, *The Historie of Serpents; or,*
the Second Book of Living Creatures, 1658

Dragons have lurked in the dark recesses of human imagination throughout our evolutionary history. They are among the most universally shared cultural legends, from Mesoamerica to Nordic Europe, Babylonia, and the Far East, generating considerable debate and discussion on the genesis and evolution of dragons, their seemingly universal presence in human mythology, and their subtle or not so subtle variations in time and place. In their most basic form dragons around the world are recognizable as chimeras with a fundamental reptilian body plan, harboring impressive powers that can be used for good or for bad. The *Oxford English Dictionary* defines the dragon as a mythical monster, represented as a huge and terrible reptile, usually combining ophidian and crocodilian structure, with strong claws, like a beast or bird of prey with a scaly skin; frequently represented with wings, and sometimes as breathing out fire. In the human consciousness, dragons are strongly linked with another mythical reptile, the serpent; in fact the English word "dragon" derives from the Latin term *drakon*, referring to a large snake. Dragons and serpents are both universally recognized as ambivalent mythical/allegorical creatures, but where did they come from? The universal tendency to create and believe in an imaginary fauna must reflect some very rudimentary human need. It is apparent that the form and sense of our mythical creatures and monsters do not spring out of thin air; they are molded representations of actual human encounters with living animals, transformed through the lens of thousands of years of human experience. The serpent is unmistakably based on the snake, simple in form but complex in its symbolism. Like serpents, dragons are recognizably reptilian, but absent the simple body plan. Dragons mean different things to different people; they are a human construct that has coevolved with us in response to our ecological and evolutionary circumstances, transformed by human nature, culture, and religion into the mythical hybrids they now embody. What are the forces at work in our psyche, our culture, and our interaction with the external world that have given rise to dragons?

The dragons of the world have a split personality, a reflection of the remarkable dichotomy between Eastern and Western dragons. In the Western world, the cultural roots of dragons can be traced back to Egypt and the Fertile Crescent. After passing through ancient Greek and Roman societies, they emerged in European myth and folklore as powerful, but basically malevolent, creatures. In the West, dragons fall into the broad category of "monsters," powerful, hybrid creatures; metaphors for what must be repudiated by the human spirit and that, in turn, create the need for heroes to defeat them in an allegory of harmony triumphing over chaos (Gilmore 2003). The Oriental dragon shares some similarities of form with its Western counterpart, likely a result of its primeval reptilian DNA, but it is worlds removed from the Western congener in its symbolism. A creature forged in the predynastic past of China, it does not play by the rules of its more notorious Western cousin. It is a different beast altogether, and the reasons behind this, including links to the Chinese alligator, are some of the issues we address in this chapter.

Tracing the origins of the Oriental dragon is not a straightforward task. The zoological pedigree of Chinese dragon mythology is obfuscated by the lack of alignment between our present, taxonomically based perceptions of species, and that of the ancient Chinese, who relied more on a societal model that interpreted animals and their behavior in the light of human experience and needs. While ancient Chinese texts contain ample references to everyday animals, they are typically in the context of their utilitarian or ritualistic uses, and almost nothing was written, or has survived, about the natural history of the fauna in question. As noted by Sterckx (2005), "Early Chinese writings rarely classify the animal world and its members as individuals and classes. None of these sources are concerned with the systematic description of animal life and morphology." So, while Chinese texts may provide some clues, they cannot be expected to provide much more than vague hints at the biological origin of dragons.

At its most basic level the Chinese dragon, like its Western counterpart, has a reptilian form; an elongated body, resembling that of a snake, but with well-defined limbs clearly brings a crocodilian to mind. But grafted onto this fundamental body form are a number of unmistakably nonreptilian features. Many analyses of the origin of dragons have focused on these "morphological" attributes and used them to link a variety of animals with Eastern dragons. But one difficulty with this approach is that there is really no standard dragon morphology, it has evolved considerably over the millennia, and at any one period of time in Chinese history there was considerably variability in how dragons were depicted. Doubtless many of the more standardized features, such as the antler/horns, or chin barbels, were taken from other recognizable species (deer, catfish), making it apparent that based on its modern appearance, the Chinese dragon is of mixed pedigree, a composite of the morphological features of a variety of flesh and blood animals. Nevertheless, the phenotype and the powers ascribed to Chinese dragons have changed over the millennia and the question remains, if we go back far enough, can we identify an animal, or animals, that served as a spark for the tradition of Chinese dragon mythology? Analysis of this topic should serve not only to explain the origins of the dragon's phenotype, or physical appearance, but also to clarify what can be referred to as the dragon's genotype, its inner nature. What is it that made the dragon image so important and so appealing to the early Chinese? In so doing we address the Chinese dragon's two most consistent features: a reptilian body form and a strong link to water.

In this chapter we present evidence that links the dragon myth with China's living "dragon," the *tu long*, or earth dragon. In the long tradition of dragon study, some have included alligators among the pantheon of Eastern fauna thought to have influenced the legend of the dragon (D. Wang 1988; Porter 1996), whereas others have rejected this idea (Q. Zhao 1992). But no one, to our knowledge, has presented a specific hypothesis concerning the timing and rationale for a particular species' role in the birth of the dragon legend. Using historical, archaeological, and paleoclimatic data, as well as what we know about the biology of the alligator, we attempt to piece together the historical puzzle that is the Chinese dragon.

SERPENTS AND DRAGONS

Dragons are a prominent subclass of serpents, the most universally recognized of mythical creatures. The serpent is instantaneously recognizable, because it is in outward appearance essentially a snake. While most mythical fauna are hybrids of various species with phantasmagorical combinations of features, the serpent is simple in form. Figures of serpents are seen in prehistoric art from around the world, and references to mythological snakes are a regular theme in the oldest written texts, ranging from Sumerian cuneiform tablets and Egyptian mortuary texts to ancient Hindu hymns (Mundkur 1978). The serpent is usually portrayed as a divine agent whose actions are venerated, but not always positive. The ambivalent symbolic nature of the serpent can likely be traced to the widespread occurrence of venomous snakes, hidden agents of suffering or sudden death, and the fact that snakes regularly shed their skin and emerge renewed and transformed, a representation of regeneration and rejuvenation. A number of mythologies have great cosmic serpents, such as the Jörmungandr, the Midgard serpent of Norse mythology that encircled the world in the ocean's abyss, or Dan, a giant serpent in West African Dahomey mythology that supports the universe on its many coils. The biblical account of the serpent in the Garden of Eden is one of the most widely recognized Christian allegories, bringing forbidden knowledge to humankind. In the New Testament, the serpent is closely associated with Satan and is a symbol of evil. In ancient Egypt, the serpent-god Nehebkau guards the entrance to the underworld. In the epic of Gilgamesh, a Sumerian text that is arguably the oldest surviving work of literature, the serpent consumes the plant of life and becomes a symbol of immortality. The rod of Asclepius, a serpent entwined around a staff, is an ancient Greek symbol linked with medicine and healing of the sick; it forms an integral part of the emblem of many health organizations, including the World Health Organization and the American Medical Association.

The mythical serpent is unmistakably a representation of the snake, and today, as in the past, snakes are among some of the most universally feared animals. The fear of snakes is a human characteristic with a complex history, deeply entwined with our evolutionary relations with snakes as an unpredictable, and in some areas significant, cause of human mortality. As argued eloquently by E. O. Wilson in his book *In Search of Nature*, serpents are the demonic dream images of flesh and blood snakes, images that symbolize our complicated relationship with nature. Even in today's world, where much of nature has been tamed, it is estimated that there are some 5 million snakebites a year resulting in 150,000 human deaths (J. White 2000). Our ancestors evolved in a world fraught with risks from dangerous animals. In an ecological and evolutionary sense, a heightened sensitivity to these potential threats would have had survival benefits. A recent study even suggests that one of the evolutionary forces driving increased human brain size and complexity was the need for better vision and quicker responses to detect dangerous cryptic predators such as large constrictors or lethally venomous snakes (Isbell 2006). Emotional reactions or a feeling of fear in response to the sight of dangerous animals would have enhanced the ability to avoid these creatures and thereby increase one's survival probabilities. This line of reasoning has been used to suggest an adaptive biological basis for a generalized fear of snakes (Öhman et al. 2001). "It pays in elementary survival to be interested in snakes and to respond emotionally to their generalized image, to go beyond ordinary caution and fear" (Wilson 1996).

But it was not only snakes that troubled the dreams and imagination of our early ancestors; our evolutionary history has primed the pump for our fascination and fear of other potentially dangerous animals. While the risk of death from snakebite has long been a feature of human evolution, early hominids developed in an evolutionary milieu that contained animals such as wolves, lions, tigers, bears, and crocodiles, all of which viewed people as potential food. Not so very long ago, human habitats harbored the reasonable expectation of encountering one or another of these dangerous creatures. And while a venomous snake strikes out at humans in self-defense, the behavior of a large, man-eating carnivore is something very different. And it was in such a landscape, one shared with creatures that could and did devour humans, that our sense of human identity arose (Quammen 2003). Today it is a rare event, but when it happens, an animal killing and eating a person attracts widespread attention. As we were writing this book, an American doctor was pulled out of a canoe by a crocodile in Botswana and killed. The news of this event was transmitted worldwide, eliciting shock and horror in millions of people. If the same doctor had drowned while swimming in the river or had been killed in a car accident, the world's press would have paid it little notice. Being killed and eaten by a predatory animal is something that we consider at a most basic level as malicious and deliberate, and it would certainly have weighed heavily on the minds of our ancestors. In the same sense that an inherent disposition to "overlearn" a fear of snakes could help keep us safe from venomous serpents, a fascination or heightened interest in predatory animals would have been an evolutionary advantage, predisposing us to better understand the habits and behavior of those creatures that sought to make a meal of us. The risk of being eaten by a hungry animal still exists, but in vanishingly small corners of the planet, and it is not even a remote concern for most of us. But a widespread interest in large predators remains with us, a ghost from our evolutionary past.

Among the species that ate early hominids in Africa is the Nile crocodile, well known even today for its man-eating proclivities (J. Ross et al. 2000). While most crocodilians today have a generally undeserved reputation as man-eaters, two species clearly warrant the title, the Nile (*Crocodylus niloticus*) and the estuarine crocodile (*C. porosus*), the latter found throughout much of the Indo-Malay region (Chapter 2). The Nile crocodile is widespread in sub-Saharan Africa, in the forests and savannas where humans evolved. The earliest known fossils of the genus *Homo* date from the late Pleistocene, approximately 2 million years ago. At that time the Nile crocodile and three other species (*Mecistops cataphractus, Crocodylus lloidi,* and *Euthecodon brumpti*) were living in the eastern and southern Africa (Tchernov 1986). Many of the fossil remains of early hominids come from river and lake deposits in regions such as the Afar Triangle and lower Omo River in Ethiopia; Olduvai Gorge, Tanzania; and Lake Turkana in Kenya, habitats that were also home to crocodiles. Not surprisingly, some of the bones of large mammals from Olduvai have crocodile teeth marks (Njau and Blumenschine 2006), and it is likely that humans were among the mammalian prey items taken by crocodiles in these areas. For people living around water bodies in Africa, being grabbed by a crocodile was and still is a very real fear, one that may have left the indelible imprint of large reptilian shapes in our subconscious.

But why is there such a universal tendency to go beyond the reality of living predators and create even more terrible and powerful hybrid creatures such as the Leviathan of the Bible, sea serpents, griffins, gargoyles, chimeras, and hydras? The propensity of our species to create these hybrid creatures is so universal it must reflect a fundamental aspect of human nature. In some cases these mythical creatures were associated with deities whose powers were symbolized by the special characteristics of the animals used in the amalgam (Hornblower 1933). Creation of these dangerous hybrids may also be linked to the use of animals as totems, which was an important aspect of human cultures in different parts of the world. In the early Zhou Dynasty in China, different tribes were represented on flags by totemic animals, and one of the many theories concerning the development of the Chinese dragon is that in warfare the totem of the victorious tribe would take on some of the characteristics of the vanquished foe (Q. Zhao 1992).

The use of real or imaginary creatures as totems, however, cannot begin to explain the amazing diversity of imaginary creatures; there must be other forces at work. Wilson (1996) suggests that people's inadequacy in grasping and processing reality in all its diversity results in a reliance on symbols, many of which have emerged from the jumbled images in our dreams. In some sense, it may also have been an evolutionary advantage to conjure up these symbolic threats to our well-being even more heinous than the flesh and blood animals from which they were derived. In his account of modern human relations with predatory animals, Quammen (2003) notes that spiritual beliefs were a means of coping with the unpredictable dangers of the world in which early humans lived. In this sense it is our culture, the totality of socially transmitted behavior patterns, arts, and beliefs, that is the link between everyday experiences and the spirit world. Wilson also argues that it is our culture that transforms the snake into the serpent. A similar line of reasoning may apply to the human need to transform a collection of dangerous, predatory beasts into mythical superpredators. One analysis of the evolutionary basis of dragons argues that for early humans, fear and anxiety can be interpreted as adaptive responses to predators and that dragons represent an amalgam of creatures that would prey on or kill primates and hominids (Jones 2000). This argument is built on just three groups of animals: snakes, predatory birds, and big cats. We suggest that this limited cast of predators should be expanded to include other actual fauna that have played important roles in the creation of the hybrid dragon creature. Particularly notable are the strong historical links between crocodilians and dragons.

The chimerical nature of dragons can also be examined within the context of monsters, another set of human constructs that transcends cultural boundaries (Gilmore 2003). Monsters are a collection of legendary creatures that harm or kill people. Much of the inventive nature of humans can be seen in the seemingly endless cast of monsters we have created, evident from the earliest writings in Mesopotamia to the current fare at the local multiplex. We need monsters, at the very least as a foil for our heroes. Monsters represent all that is beyond human control and are a device upon which we can hang our deep-seated worries of unmanageable, chaotic nature. They are constantly being reinvented as novel and ever more dangerous creatures, but they cannot be constructed from nothing, they require a basis in our human experience and evolutionary history. Their features are derived from nature by reshuffling the characteristics of dangerous, predatory, and fearsome animals into even more horrid composite creatures.

EASTERN VERSUS WESTERN DRAGONS

Western dragons fit the monster mold, Eastern dragons do not. While outwardly similar in appearance, these two dragons are fundamentally dissimilar in their temperament; they are similar in their phenotype, but their underlying genotypes are quite distinct. Chinese dragons are usually represented as a multiple chimera—an amalgam of several different animals, frequently including a snake, deer (the antlers), fish, tiger, and ox (Bates 2002). Western, or European, dragons are customarily depicted with wings (which Oriental dragons, although frequently shown as flying, do not possess) and breathing fire. But there is a fundamental difference in the nature of Eastern and Western dragons: The former is generally viewed as a force for good and the latter, evil.

The origins of dragons in western Asia and Europe extend back to the earliest human civilizations that grew in the Fertile Crescent and the lower Nile, while the Chinese dragon traces its ancestry back to East and Southeast Asia. The different geographic origins of the two types of dragon may be one of the reasons why they are so unlike. In part, the dragon dichotomy may reflect dissimilar human experiences with the animal models, including crocodilians, that influenced the development of the dragon. In China, we postulate that it was the Chinese alligator, a small and relatively innocuous species, that provided the seed for dragon beliefs. The legacy of the European dragon includes the large man-eating Nile crocodile among its ancestors. The Chinese alligator as an early model for the Chinese dragon may also explain its beneficent role as a water deity.

In the West, some of the earliest references to dragonlike creatures come from the valleys of the Tigris and Euphrates rivers. Mesopotamian mythology refers to several dragonlike gods with features that embody the special qualities of their parental species such as "the fierce strength of the lion, the keen swiftness of the eagle, or the mystery and deadliness of the snake" (Hornblower 1933). The most important of the serpentlike deities was Tiamat, who figures prominently in the Babylonian tale of creation. Tiamat, an angry goddess who wished to destroy the other gods, was vanquished by Marduk, the city god of Babylon. Marduk agrees to kill Tiamat if he is made supreme god, and after doing so he creates heaven and earth by cutting her body in half, with the Tigris and Euphrates rivers flowing from her eyes. The legend of Tiamat may have been based on even earlier Sumerian myths of the dragonlike creature Zu, the representation of watery chaos (Q. Zhao 1992). The extent to which crocodilians may have influenced dragon myths in Mesopotamia is not well known. Today there are no crocodilians in either river. Just to the east, southeastern Iran is the limit of the distribution of the mugger crocodile (*C. palustris*). Historically, Nile crocodiles were found in what is today Israel, even in the River Jordan (Anderson 1898). However, although the Tigris and Euphrates rivers would seem to offer good habitat for either of these species, there is no historical or archeological evidence that either was ever found there.

The story is very different for the lower Nile, the cradle of Egyptian civilization. The ancient Egyptians shared the entire length of the lower Nile, from Nubia to the delta, with large numbers of crocodiles that, along different parts of the river, were either worshiped or demonized. Among the pantheon of Egyptian gods, Sobek was represented in the form of a crocodile. Crocodiles preyed on the river dwellers, and for this they were feared. Fishermen were known to conjure magic spells or sing holy songs to protect themselves while crossing the river. But as embodied in Sobek, crocodiles were also worshiped as a deity who controlled the ebb and flow of the Nile and was known as "he who causes to be fertile." Farmers believed that they could tell when the waters of the Nile would start to drop by when the crocodiles nested. The Egyptian god Set (or Seth) is considered to have been among the most powerful of Egyptian deities. Usually represented as a dragon-like composite of several animals, including the crocodile and hippopotamus, Set came to represent evil (Bates 2002). Strong Egyptian-Greek links emerged as early as the sixth century BC, and the legend of Set and Sobek may have had a prominent role in the emergence of the dragonlike Greek creature Typhon, which was reptilian in form and associated with winds and volcanic eruptions (Q. Zhao 1992). Leviathan, the Old Testament monster described in the Book of Job, is widely considered to be based to a large extent on crocodiles (Q. Zhao 1992), a belief that can be traced back to 1663 in Bochart's *Hierozoicon*, a work that deals with all the animals mentioned in the Bible and their treatment by ancient naturalists.

In Europe stories of dragons proliferated and their study was undertaken by the most renowned scientists of the time. In AD 23, the Roman naturalist Pliny the Elder provided the first written account (*Naturalis Historia*) of the fauna of regions beyond Europe. Many of his stories were based on hearsay with a hefty dose of imagination, but his work remained the standard natural history text until Konrad Gesner, a Swiss naturalist, published his three-volume *Historia Animalium* (1555–1558), which is frequently considered to mark the beginning of modern zoological studies. A fourth volume, on serpents, was printed posthumously and subsequently used as the basis for *The Historie of Serpents* published in English by Edward Topsell in 1608. Besides accounts on snakes, lizards, and crocodiles this compilation included extensive descriptions of dragons.

European mythologies frequently used dragons as the embodiment of evil that is symbolically vanquished, perhaps influenced by some of the earliest Sumerian legends. In northern Europe a dragon figures prominently in the Anglo-Saxon legend of Beowulf written in the eighth century. As a youth Beowulf killed Grendel, a fearsome monster that was attacking the hall of the Danish king Hrothgar. Beowulf dies some 50 years later when, in a prideful attempt to protect his people, he is mortally wounded by a dragon. In the Volsung saga, a thirteenth-century Icelandic narrative based on older Norse legends, the hero Sigurd kills the greedy dragonlike Fafnir and eats his heart. This story parallels that of the Nibelungenlied, an epic poem in Middle High German about the dragon-slayer Siegfried at the court of the Burgundians. As Christianity spread throughout Europe, Western dragons became largely a product of Judeo-Christian reinterpretations of older legends, appearing as monsters in literary works and as religious symbols of the victory of Christianity over paganism. The story of Saint George and the dragon is the iconic representation of this reinterpretation (Box 4.1). The legend of Saint George and the dragon may have had its basis in some ancient conflict with a real crocodile, and in some of the images of the battle the dragon has crocodile-like features. Many of the subsequent legends of killing dragons in Europe appear to be based on crocodiles, which were at one time at least occasionally encountered around the Mediterranean. In France churches at Marseilles, Lyons, Ragusa, and Cimiers, exhibit crocodile skins or stuffed crocodiles as the remains of dragons (Chambers 1881); this is also true in Italy, Switzerland and Spain.

CHINESE DRAGONS

When the first Europeans arrived in the Far East they encountered legends about a large, powerful, reptile-like creature that shared many features of the European dragon. Unlike the malevolent European dragon, the Chinese dragon, *long*, represented strength and goodness and came to symbolize Chinese royalty. Although its outward appearance resembles that of a reptile, the *long* was usually imbued with the power of flight, even though (unlike the Western dragon) it had no wings. Chinese dragons are believed to live underwater or underground, are usually associated with rivers or other wetlands, and control thunder, clouds, rain, and flooding. This link to water is one of oldest and most fundamental aspects of the nature of the Chinese dragon. Many modern depictions of Chinese dragons show them with a pearl of wisdom, which is frequently interpreted as the source of their supernatural powers. Today Chinese dragons are usually associated with good fortune, generosity, and wisdom. In the Taoist Chinese cosmology, dragons represent the force of yang; the phoenix, another species in the Chinese legendary fauna, represents yin. These two complementary forces, yin and yang, constitute all aspects of life. Dragons are also powerful cultural icons in Korea and Japan, as well as parts of Southeast and South Asia including Vietnam.

Some attempts to explain the form of the Chinese dragon are linked to Huang Di, the Yellow Emperor of predynastic Neolithic China. During this period, some 6,000 to 7,000 years BP, what is now northern China was peopled by a variety of tribes, or clans, centered on the Yellow River valley. Clans were represented by a specific totem animal, a symbol of good fortune, which was depicted on their banners during battles. Huang Di, who is most likely a historical amalgam of many tribal chiefs, is credited with uniting these tribes under one banner by fusing together the many tribal totems that included bears, wolves, tigers, leopards, snakes, and birds of prey. The fusion of these totem animals into one, unified representation has been suggested as the genesis of the dragon as an amalgam of many living species (Q. Zhao 1992). But this interpretation of the Chinese dragon's origins is unsatisfactory as

BOX 4.1

The Legend of Saint George and the Dragon

Saint George, one of the most widely recognized Christian martyrs, is renowned for his defense of those in need. He is the patron saint of England, a land he never visited. The symbol of Saint George, a red cross on a white background, became the flag of England (not to be confused with Great Britain's flag), but the most iconic images of Saint George are those that show him killing a dragon. While most of the life of the historical figure upon which Saint George is based is shrouded in mystery, he was probably born in Cappadocia, a region in modern Turkey, of noble, Christian parents. Following the death of his father, he traveled with his mother to her native Palestine where she owned land that George may have been expected to oversee. Instead, he entered the Roman army, where he rose to hold an important post—the rank of tribune (equivalent to a colonel in modern terms). Beginning in AD 302, the Roman emperor Diocletian severely persecuted Christians, and Saint George complained personally to the emperor about the appalling purges. He was thrown into prison, tortured for refusing to recant his faith, and then dragged through the streets in Lydda, Palestine, and beheaded.

Many of the legends concerning Saint George are derived from the fictional accounts of his life in the *Acta Sancti Georgii* (Acts of Saint George), which were written at a very early date and have been embraced by the Eastern Church since the fifth century. The fame of Saint George spread throughout Europe with the 1265 publication of the *Legenda Sanctorum* (Legends of the Saints), later known as the *Legenda Aurea* (The Golden Legends), by Jacobus de Voragine. This was the book that popularized the legend of Saint George and the dragon, a story that struck a receptive chord in England because of similar accounts of dragons in Anglo-Saxon lore.

The origin of the Saint George and the dragon fable remains somewhat obscure (Hogarth 1980). The slaying of dragons was a common theme in the Near East from the time of the ancient Sumerians, whose myths included Marduk killing Tiamat. There is some evidence that the legend of Saint George may have had its origins in an ancient Egyptian account of the god Horus killing a crocodile (Clermont-Ganneau 1877). In the story as we know it today, a dragon lived in a lake near Silena (possibly a misspelling of Cyrene, an oasis settlement in central Libya or possibly of the name of a small village [Zilin or Selen] on the coast of Tripolitania, near Lepcis Magna). Whole armies were said to have gone up against this creature without being able to kill it. The dragon ate two sheep a day, and when they were scarce, lots were drawn in local villages and young women were sacrificed to the dragon.

The iconic Saint George slaying a dragon. From a cabinet painting by the Italian Renaissance master Raphael (1504–1506).

George happened to arrive in the area on the day when the daughter of the king was to be sacrificed. According to the legend George agreed to slay the dragon in exchange for a promise by the king's subjects to be baptized. He then crossed himself, rode to battle against the creature, and killed it with a single blow of his lance. In some versions he merely incapacitates the dragon, which is then led into town by the princess and beheaded. In Christian mythology the slaying of the dragon is largely viewed either as an allegory of Emperor Diocletian's persecution of the Christians or more generally of Christianity's victory over paganism.

Figure 4.1. Various designs of Chinese dragons.

it does little to explain the fundamental link between the dragon and its essential element, water. As noted by G. E. Smith (1922): "Wherever the dragon is found it displays a special partiality for water." It is this link with water that is, within the context of the emerging Chinese civilization, the key to understanding the Oriental dragon.

The civilization that developed on the East China Plain is among the oldest on earth, and like so many societies around the world it grew out of an expanding agrarian tradition that allowed people to settle in villages and enjoy, at least periodically, a surplus of food. During the last glacial maximum in the late Pleistocene, much of China was a cold, dry plain inhabited by bands of Paleolithic hunter-gatherers. During the transition from the Pleistocene to the Holocene the climate warmed, rainfall increased, and the stage was set for early attempts at agriculture. Archeological evidence suggests that approximately 13,000 years ago much of the Pleistocene megafauna, on which the hunters had depended for food, died off and the previously nomadic groups began supplementing their diet by harvesting rice and millet (T. Chang 1987). Two apparently independent agricultural centers developed in Neolithic China, one in the Yangtze River basin based on the cultivation of rice, and another along the more northern Yellow River founded principally on the cultivation of two kinds of millet (T. Lu 1999, 2005). While water, whether flowing in rivers or falling as rain, was essential in the development of the Neolithic agricultural cultures in both regions, it was likely a more critical factor in arid northern China. Being able to predict rainfall patterns, or even better, to control its availability, was a matter of life and death. But while it was essential for life, water can also be a destructive force, and this is perhaps nowhere better illustrated than by the Yellow River. It is frequently referred to as "China's sorrow" because of its long history of flooding and shifting course, which has resulted in the deaths of millions of people. Water, with its life-giving and life-taking properties, was both a blessing and a curse for Neolithic farmers (Q. Zhao 1992).

The history of China is replete with ambivalent notions concerning water. From the legendary, pre-dynastic period of Huang Di emerged the earliest semi-historical periods, the Xia (ca. 2200 BC) and the Shang (ca. 1700–1100 BC) dynasties. The Xia Dynasty was founded by Yu the Great, a semihistorical figure who was the first to be associated with the control of water and the integration of water management programs, which were essential in the development of a Chinese State (Q. Zhao 1992). Yu's father, Gun, was assigned to control the devastating flooding of the Yellow River. Gun's attempts to control flooding, which were based on strengthening the levees, were unsuccessful and the task passed to Yu. Yu's approach was different; he dredged the river. And after 13 years using the labor of thousands, Yu was successful. The emperor was so impressed with Yu's accomplishments that he passed the throne to him instead of his own son. In Chinese legend, history, and geography, the tale of Yu is interwoven with that of dragons that control drought and flooding for the benefit of mankind. Today more than 40 major rivers in China are named after dragons.

In a translation of the Chinese classic *The Works of Mencius* made by James Legge in 1861, book 3, part 2 relates the account of Yu (Yaou), who was in the employ of Emperor Shun, battling the waters and using the presence of snakes and dragons as a metaphor for the destructiveness of the flooding:

In the time of Yaou, the waters, flowing out of their channels, inundated the Middle kingdom. Snakes and dragons occupied it, and the people had no place where they could settle themselves. In the low grounds they made nests for themselves, and in the high grounds they made caves. It is said in the Book of History, "The waters in their wild course warned me." Those "waters in their wild course" were the waters of the great inundation. Shun employed Yu to reduce the waters to order. Yu dug open their obstructed channels, and conducted them to the sea. He drove away the snakes and dragons, and forced them into the grassy marshes. On this, the waters pursued their course through the country, even the waters of the Keang [Yangtze], the Hwae [Huai], the Ho [Yellow River], and the Han, and the dangers and obstructions which they had occasioned were removed. The birds and beasts which had injured the people also disappeared, and after this men found the plains available for them, and occupied them.

Yu the Great's ability to combine the skills needed for water control and for national leadership is among the earliest accounts linking together Chinese royalty and dragons (Q. Zhao 1992), a bond that continued through history. The first emperor of the Western Han Dynasty, Bang Liu (206–195 BC), was said to have been "seed" of the dragon because his mother dreamt that she had been impregnated by one. But not until the Song Dynasty (AD 960–1279) did dragons come to be used as royal emblems, based in part on the belief that they rose from earth to heaven—something that the Song Dynasty emperors thought they, but not the common people, could do (Bates 2002). Dragon images were used on royal clothing, banners, and even as part of the architecture in the construction of the emperor's dwellings. Hongwu, the first Ming Dynasty emperor, further refined the use of dragons as a royal symbol and standardized the features of what we know today as the Chinese dragon; many other dragon forms were relegated to the status of "sons" of the dragon (Bates 2002). The Forbidden City in Beijing, home to Chinese royalty during the Ming and Yuan dynasties, is covered in dragons, with 13,844 images decorating the Hall of Supreme Harmony alone (Raffaele 2008).

The strong links between Oriental dragons and water are also evident in the dragon's role as rain deity. The ancient Chinese believed that rains came when sleeping dragons emerged from bodies of water and rose to heaven (Marks 1998), and there are many historical references to dragons either accompanying or appearing immediately after heavy rains. Floods were believed to be the result of dragons fighting in rivers or in the air (de Visser 1913). *Comments by Zuo*, an interpretation of the *Spring and Autumn Annals* by Confucius, notes the appearance of dragons during a flood in the nineteenth year of the Zhaogong emperor (523 BC), when "the dragons were fighting in the deep pools of the Wei River outside the Shi Gate" (Bates 2002). Chinese dragons were thought to be able to move clouds and bring rain, and for many centuries people would appeal to dragons for rain during periods of drought. In what is today southern China, dragons and tigers were considered to be the yin and yang of rain and drought. During droughts people would drag tiger skulls through rivers to awaken the dragons and bring rain (Marks 1998). The Chinese classics contain references to people throwing things that angered dragons, including poisonous plants, stones, or tiger bones, into their pools in an attempt to make rain (de Visser 1913). In Beijing during the Qing Dynasty (1644–1911) people believed that rain could be produced by throwing iron into the Pool of the Black Dragon in an attempt to frighten the dragon and spur it on to make rain (Bates 2002).

On the plains of northern China where the rains have long been unpredictable, prayers to dragon rain deities were made to guarantee water for crops. Rain rituals as early as the sixth century BC involved a dragon image animated by a procession of dancers. Bone carvings of dragons from the third century BC have a reptilian, alligator-like body with heads that extend into clouds and are thought to have been used by diviners predicting rainfall patterns (Hopkins 1913). Other physical representations of dragons were used to induce dragons to bring rain. In the Han, Eastern Han, Tang, Song, and Qing dynasties, people used clay dragon images during times of drought, either "sacrificing" them or taking the images into the fields to show them the effects of the drought. Another prescription for rainmaking cited in the Chinese classics, placing "water lizards" in water-filled jugs with wooden covers, has a seemingly direct link to alligators., These jugs were then to be incessantly struck, day and night, by 10 young boys using green bamboo sticks until it rained (de Visser 1913). A noteworthy point here is that in the early translations of the classics the character for *T'o* was often interpreted to mean a lizard (Fauvel 1879). The same unusual account of rainmaking cited by de Visser (1913) also notes "that dragons and water lizards belong to the same species." The image of dragons remains closely tied to that of water even in modern-day China where a water faucet is referred to as a "dragon's head" and fire hoses are "water dragons" (Bates 2002).

Alligators are not the only reptile to have influenced dragon mythology. Snakes are also part of the mix. Buddhism probably entered China in the first century BC, and after the Jin Dynasty (AD 265–420), the belief in Oriental dragons as water deities was strengthened by the Buddhist notion of nāgas. In ancient India nāgas were serpent folk deities associated with water, and through this power they were responsible for endowing kings with their authority (Bloss 1973). These serpent deities were subsumed into the Hindu and Buddhist religions, where the nāgas are important icons with strong connections to water. Subsequently they may have been incorporated into the Eastern concepts and imagery of dragons.

ALLIGATORS AS DRAGONS

For alligators to have played a role in the genesis of dragon mythology, they should have been found in Central China during the early Neolithic period when Chinese civilization was emerging. While there are no crocodilians to the north of the Yangtze River basin today, that was not the case in past millennia (Chapter 3). Remains of alligators found at Neolithic archaeological sites in North Central China are much farther north than the current distribution of the species. Some of the remains are jumbled together with those of other species that were on the Neolithic menu, indicating that alligators were used for food. However, remains of alligators have also been found associated with human burials at some of the earliest known Neolithic archaeological sites on the North China Plain, along with inscribed turtle shells, pottery, and carved jade, suggesting that people not only ate alligators but also imbued them with a special cultural significance.

Among the most significant remains found at Neolithic burial sites in several parts of the North China Plain are drums made partially with alligator skin. Written

Figure 4.2. The reconstructed evolution of the Chinese character for dragon from early pictograms resembling a crocodilian. From He 1986.

records from the Han Dynasty (206 BC–AD 220) referring to the earlier Xia period (ca. 2200 BC) note that alligators were caught during the second month (early spring) and that their skin was used to make drums (Sterckx 2000). The earliest culture known to have used alligator drums was the Dawenkou, a group of contemporaneous communities living in Shandong, Shanxi, Anhui, Henan, and Jiangsu provinces approximately 4,600 to 6,100 years BP. Dawenkou burial sites in Shandong contain artifacts made from elephant ivory, as well as alligator bones and alligator skin drums (Underhill 1997). The presence of these drums in burial chambers is indicative of high social status of the deceased as they were always found in the most lavishly decorated tombs (W. Shao 2000; L. Liu 2000).

The Longshan people, a later Neolithic culture along the lower Yellow River basin (ca. 4,000–5,000 years BP) were adept at making pottery and were one of the earliest cultures to transition to larger settlements, including some surrounded by walls or moats. Wooden drums fitted with alligator skins have been found at three Longshan culture sites in Shandong and Shanxi provinces (L. Liu 2000). Remains from a cemetery at Taosi (Xiangfen County, Shanxi Province), the largest of the walled Longshan settlements, include meter-long wooden drums fitted with alligator skins. Drums fashioned with alligator skin have also been found from royal tombs of the much later Shang Dynasty (ca. 1700–1100 BC) in Anyang, Henan Province, where they are also symbols of the elite status of the deceased (Pearson and Underhill 1987). The presence of alligator skin drums in tombs ranging from the Neolithic into historical times suggests that the drums were a symbol of power and authority over a period of several thousand years in ancient China (L. Liu 2000).

Further clues to the dragon's origin come from some of the earliest known styles of Chinese writing. The Chinese character for dragon, or *long*, made its first appearance in the Shang Dynasty (Q. Zhao 1992), arguably the most important formative period in the development of Chinese civilization. It was during the Shang Dynasty that a Chinese system of writing first developed. Inscriptions used for divination on "oracle bones," the plastra of turtles or the scapulae of cattle (Chapter 3), have been found in large quantities in the ruins of Yin, the late Shang Dynasty capital (Hsu

2002). The Shang script was highly pictographic and had no standardized way to depict dragons, using some 70 variations for the character *long* (Q. Zhao 1992). Nevertheless, it is perhaps significant that some of the earliest known characters for *long* bore a resemblance to crocodilians (Figure 4.2; He 1986). Names used for dragons and crocodilians in China also frequently overlapped. In addition to *long*, dragons have been referred to as *jiao* or *jiaolong*, the former also being a term for crocodiles (Q. Zhao 1992). Even in present-day Anhui, Chinese alligators are still referred to as earth or river dragons (*tu long, he long*) by local farmers. The Shang was followed by the Zhou Dynasty (1122–256 BC), and bone carvings of dragons from this period used by diviners to predict rain were "based on the physical structure of the alligator" according to Hopkins (1913).

Although Chinese dragons have been depicted in a wide variety of ways, they nearly always have an unmistakable reptilian form (Figure 4.3). Attempts to standardize the appearance of the Chinese dragon were linked to its use as a royal symbol in the Song and Ming dynasties. Prior to this, there was a great deal of inconsistency

Figure 4.3. Song Dynasty "crocodilian form" dragon. From Bates 2002.

Figure 4.4. Neolithic burial site with images of tiger and dragon, Xishuipo, Puyang County, Henan Province. From K-C. Chang and Xu 2005.

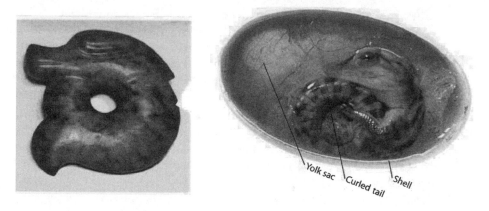

Yolk sac   Curled tail   Shell

Figure 4.5. Jade "pig dragon" (*left*) and embryo of a Nile crocodile (*right*).

in how dragons were depicted. Nevertheless, some of the earliest known representations of dragons do bear a resemblance to crocodilians. The earliest known physical
representation of a dragon in China was unearthed in December 1987 during excavations for a Yellow River diversion project in Puyang, northern Henan Province. A
burial site at Xishuipo, Henan Province, from the Neolithic Yangshao culture estimated to be about 6,000 years old, contained a human skeleton flanked by representations of a tiger and a dragon made from the shells of freshwater mussels (Figure
4.4). The dragon figure bears a resemblance to a crocodilian. Crocodilian-like dragon
images are also known from Shang Dynasty tombs at Taouhuazhuang, Shilou, and
Shangxi, as well as tombs from the Western Zhou period (ca. 1027–771 BC) (Bates
2002).

Some of the earliest known representations of dragons are small, circular jade
objects that have been referred to as "pig dragons," due to their resemblance to a pig.
Another possible interpretation is that these "dragons" are actually hatchling alligators, coiled up inside an egg (Figure 4.5).

The Chinese dragon, while obviously an amalgam of many creatures has a basic
four-legged reptilian body form. We know that alligators were found in North Central China during the Neolithic, and there is evidence that their skins were symbols
of high social status. There are also interesting resemblances between some early
dragon symbols, pictographic characters, and crocodilians. While these are all tantalizing clues, all they amount to is circumstantial support for a phenotypic alligator-
dragon link. What is needed is support for a connection between alligators and the
fundamental nature of dragons as a water deity. It is here that the biology of alligators
provides some important clues.

The earliest quasi-historical references to dragons in China are from the Chinese
classical texts, which are pre–Qin Dynasty (221–206 BC), especially the Confucian
Four Books and *Five Classics*. The most widely read of the *Five Classics* is the *Yijing* (or
I Ching Book of Changes). Tradition has it that the *Yijing* was written by the legendary Fu Hsi (ca. 2900 BC), the first of the mythical three emperors of ancient China,
but it is possible that its roots extend back much earlier. The *Yijing* lays out a system

of divination that replaced the previous system of oracle bones; it is based on interpretations of randomly selected sets of broken and unbroken lines. This system of
cosmology and philosophy was at the heart of ancient Chinese Confucianism. Some
references to the behavior of dragons in the classics could be interpreted as descriptions of alligators, such as the habit of digging burrows, which is well known among
Chinese alligators. One note on dragons in the *Yijing* sounds like the winter hibernating behavior of alligators: "The dragon … is a water animal, akin to the snake,
sleeping in pools during winter and arising in spring. When autumn comes with its
dry weather, the dragon descends and dives into the water to remain there till spring
arrives again." An appendix to the *Yijing* notes that "the hibernating of dragons and
snakes is done in order to preserve their bodies." And while the classics contain references to "water lizards" being used in rainmaking rituals, they also suggest another
potential alligator-dragon connection based on the vocalizations of alligators and
their link to the life-giving springtime rains (cited in de Visser 1913).

Crocodilians are highly vocal reptiles. Males, as well as the females of some species, produce a variety of sounds that play an important social role. The loudest
of these sounds are referred to as bellows. Alligators are among the most vocal of
the crocodilians, and American alligators produce loud, sonorous bellows that are
important in territorial defense by males and for bringing the two sexes together
for courtship and mating (Garrick and Lang 1977; Garrick et al. 1978). The annual
peak of bellowing takes place during the spring courtship and mating season. Bellowing is not only a vocal signal, but a visual one as well. Alligators bellow from a
characteristic body posture with both the head and the tail raised out of the water.
Chinese alligators follow the same seasonal behavioral pattern. And while both sexes
bellow, their calls are a series of much shorter, staccato vocalizations very different

Figure 4.6. William Bartram's drawing of American alligators on the St. John's River in Florida.
Bartram mistook the water vapor emerging from their nostrils for smoke.

Figure 4.7. Water vapor issues from the mouth of an American alligator bellowing on a cool morning. Chinese dragons were believed to breathe out moisture-laden mists that caused rain. Photograph by Kent Vliet.

from those of the American alligator. For both American and Chinese alligators, bellowing is principally an early morning activity, and on a cool, humid morning the air that is forced out of the alligator's nostrils may look like smoke (Figure 4.6). The American naturalist William Bartram wrote a much-maligned account of smoke-breathing American alligators from the St. Johns River in Florida (Bartram 1791). The "smoke" Bartram observed was likely water-laden mist emanating from the lungs of a bellowing alligator (K. A. Vliet, pers. comm.; Figure 4.7). Alligators also bellow in response to loud, low-frequency sounds, such as thunder, which can elicit a chorus of vocalizations from alligators. (Chapter 5 describes alligator bellowing behavior in more detail.)

The link between alligator vocalizations, clouds, and rain provides a strong connection to the mythical dragon. In the ancient Chinese medical work *Pen Tsao Kang Mu* (*The Great Herbal*; see Chapter 3), the alligator ('T'o) is said to make clouds by exhaling and causing rain (de Visser 1913). A section in the *T'ai-p'ing kuang-chi*, a comprehensive compilation of records from the early eleventh century, refers to alligators as "that which calls thunder." When the *T'o* cried, farmers knew that rain was coming (Fauvel 1879), a belief that persists today in areas where alligators still live in China.

Some of the earliest Chinese writing, the classics, notes the ability of dragons to produce clouds: "When rain is to be expected, the dragons scream and their voices are like the sound made by striking copper basins. Their saliva can produce all kinds of perfume. Their breath becomes clouds, and on the other hand they avail themselves of the clouds in order to cover their bodies" (de Visser 1913). Garrick (1975) likens the sound made by the bellow of a Chinese alligator to that produced by "striking the closed end of a standard sized metal garbage pail," a twentieth-century descriptive equivalent to striking a copper basin. The reference to the sweet-smelling saliva of the dragon suggests the pungent musk that is produced from the mandibular glands of crocodilians (including the Chinese alligator) and that, interestingly enough, has been used as a base for the production of perfume in the twentieth century (Medem 1983).

A case can also be made for a link between alligator bellows and the sound of thunder. Spring thunderstorms were believed by the ancients to be the work of dragons, and the *Yijing* states that "the dragon is thunder." The sharp, staccato nature of the bellow of the Chinese alligator is not unlike the noise produced by a thunderclap. In fact the similarity between the sound of an alligator bellow and thunder may have influenced Chinese beliefs in a god of thunder, sometimes illustrated as a dragon, which was thought to be a "drum of heaven and earth" (Porter 1996). The skins of alligators were used to make drums, which were a symbol of elevated social status and power. According to ancient texts, drums were associated with shamanistic rituals and were used as important ceremonial instruments (L. Liu 2000). The *Shih ching*, or Book of Odes, the first anthology of Chinese poetry compiled by Confucius (551–479 BC), notes that the sound of a drum made from the skin of an alligator was meant to resemble the call of the alligator itself (Porter 1996). Other early Chinese texts claim that alligators will call, or bellow, in response to the sound of the drum or when the alligator wants rain (Sterckx 2000). The early word for the Chinese alligator, *T'o*, was an imitation of the explosive sound of the alligator's bellow, which Fauvel (1878) noted as being "certainly an example of onomatopoeia as anybody who has heard the cry of the Alligator can testify."

The alligator-rain link would certainly have caught the attention of early farmers battling the elements to raise their crops. Farmers around the world are frequently viewed as a superstitious lot, but they are also keen observers of natural phenomena, particularly as it relates to weather and crop growth. Even among modern-day farmers in the vanishingly few areas where wild alligators remain the bellowing of alligators is considered to be a sign of changing weather more reliable than the national weather reports (D. Wang 1988). And it was during the spring alligator courtship season, when bellowing is most prevalent, that Neolithic Chinese farmers would have begun planting with the hopes that spring rains would provide the critically needed moisture for their crops. Our studies of Chinese alligators have shown that there actually is a relationship between alligator bellowing and weather, with alligators being more likely to bellow when it is rainy or overcast than when it is clear and sunny (Chapter 5). A similar pattern of increased bellowing during rain has been noted in spectacled caimans, a South American crocodilian in the same family as the Chinese alligator (J. Thorbjarnarson, pers. obs.). The relationship between alligator bellowing and rain, is, we believe, the seed that eventually gave birth to the dragon legend among the Neolithic farming societies of eastern China.

Patterns of rainfall in China, as elsewhere, are strongly influenced by terrain. Eastern China's climate is monsoonal. Summer rains are the result of land heating faster than the ocean and drawing in moisture-laden air west and northward off of the South China Sea. This seasonal movement of air masses first crosses over southern China, drawing off much of the moisture before it reaches the North China Plain, resulting in a dryer and more variable pattern of rainfall. While rainfall patterns in the North China Plain are variable today, paleoclimate studies indicate that early mid-Holocene climates were considerably more variable in Eastern China (Tao et al. 2006).

Our historical scenario for the emergence of the dragon is as follows. During the last glacial maxima, some 18,000 years ago, what is now East Central China was too cold for alligators, which survived, along with other subtropical fauna, in present-day southern China (Chapter 3). During the early Holocene, there was a period of rapid planetary warming and sea level rise, and the Yangtze region was once again a hospitable habitat for alligators. Some 5,000 to 6,000 years ago temperatures were warmer than they are today, and alligators extended their range up into the Huai River drainage and possibly into the lower Yellow River basin. At the same time there was a rapid evolution in human societies. Much of the Pleistocene megafauna, such as mammoths, woolly rhinoceros, and other species upon which humans preyed, disappeared from the scene. Hunter-gatherer societies had likely long collected seeds of certain grasses for food and could even store them for lean times. The transition to agricultural societies began when people started planting seeds, a process that eventually led to the development of domesticated versions of wild rice and millet, the two principal grass crops in early Chinese history. Rice requires water for germination and growth and so rice growing societies would have settled along the margins of lakes, rivers and marshes; areas where it is likely they would have been intimately acquainted with the Chinese alligator. Because of its size, the alligator was not regarded by people as a dangerous predator. And there is archaeological evidence that they were used as food (Chapter 3). Keen-eyed hunter-gatherers and early farmers who came across alligators in the spring as rains and life were returning to the once wintry landscape would have been familiar with their bellowing rituals. Here was a creature that lived in the water, disappeared during the winter, reappeared with the warm temperatures of the spring, and whose calls seemingly brought clouds and rain.

Rain was the lifeblood of farmers, who likely would have placed special significance on the relationship between an alligator's bellowing and the coming of rain. These early cultivators could not have failed to notice the resemblance between the alligator's calls and the sounds of thunder, and would also have observed that on cool mornings vapor, or clouds, issued forth from the nostrils of alligators as they called. From these observation could have sprung the belief that alligators possessed special powers, and that their springtime bellowing was to ensure the rain they needed to create their wetlands home by flooding low lying areas. Their disappearance in the fall could have been seen as a harbinger of the dry, cold months ahead. When times were dry and the rain elusive, drums made from the skin of alligators could have been used to elicit bellows from alligators and encourage them to bring the life-giving rains.

The association between farmers and alligators in the Yangtze basin would likely have been ongoing during the lengthy transition period from harvesting wild rice to cultivating a domesticated crop. The first evidence of rice gathering in the Yangtze dates to about 10,000 years BP; the scarcity of archaeological sites from this period makes it difficult to be more precise (T. Lu 2005). By 8,500 years BP, rice cultivation had spread to the Huai River valley, but there seems to have been little or no rice cultivation in the lower Yellow River until approximately 7,000 years BP (T. Lu 2005). The delay in the arrival of rice could have been the result of climate factors or the lack of early cultural diffusion from the south into the Yellow River region. Not until the Yangshao Period (ca. 7,000–5,000 years BP) do there appear to have been significant cultural exchanges between the two regions (K.-C. Chang and Xu, 2005). Bellwood (2005) suggests that this may have been a result of a Yangtze diaspora caused by population expansions related to rice cultivation, as well as the domestication of chickens and pigs. Today the Yellow River valley is north of the rice cultivation zone in China, but archaeological evidence suggests rice was grown there on and off until the Shang Dynasty (T. Lu 2005).

The Yangshao period was one of increasing cultural sophistication and social organization in Central and North China (K.-C. Chang and Xu 2005). Rice cultivation in the Yellow River valley would have been strongly influenced by cultural exchanges with people from the south. Rice must have been a very attractive crop to the Yellow River cultures as rice grains and yields are much larger than those of millet (T. Lu 2005). Nevertheless, with cooler temperatures and less consistent rainfall, rice yields in the Yellow River region were probably considerably below those from the climatically superior Yangtze valley. This would have become even more of a problem in the Yellow River after 4,500 years BP when East Asia entered an ongoing period of cooling and drying.

The juxtaposition of these factors—the less favorable climate for rice cultivation, lower yields, and cultural exchanges with Yangtze and Huai River cultures—set the stage for the evolution of the alligator as a symbol of rain for farmer, into a "protodragon" in the Neolithic agricultural communities along the Yellow River. The first known dragon images appeared during the Yangshao period, and it is perhaps significant that these were in northern, dryer regions as opposed to the Yangtze River valley. Of course rain was vitally important for the cultivation of upland crops as well, and it is possible that the alligator-protodragon came to be linked with rain outside of the low-lying lake or riverside communities. Farming cultures along the Yellow River may have been vaguely familiar with alligators; they were living in areas just to the south in the Huai River basin (Chapter 3). Alligator skin drums, with their symbolism for rain and rainmaking, were desirable articles traded from the south: icons of status and power. The Yangshao was also a period when the emerging Chinese cultures underwent profound changes in their systems of religious beliefs, including the possible emergence of a class of religious practitioners (K.-C. Chang and Xu 2005). The earliest known representation of a dragon, which, with one of a tiger, flanks buried human remains at Xishuipo (Figure 4.4), has been interpreted by some as being indicative of the deceased's role as a shaman (K.-C. Chang and Xu 2005), which may have included weather management (rainmaking). Burying someone with alligator skin drums may have been an attempt to ensure sufficient rain for the deceased in the afterlife, as well as being a status symbol.

Over time, the alligator-inspired protodragon would have evolved. The lack of familiarity with alligators in northern China may have facilitated rapid changes in the way the developing dragon images were portrayed. In the long transition from the Yangshao Late Neolithic cultures through the first of China's legendary dynasties, the Xia, Shang, and Zhou, the climate began to cool. Areas that were formerly lakes and marshes on the North China Plain dried up. The northern populations of the Chinese alligator in the Huai River basin would have disappeared, beginning a period of retreat to the warmest corners of the Yangtze region. Over an interval of one to two thousand years the protodragon morphed into the dragon, in all its multiplicity of forms, with the addition of features from a host of other species of special importance in the emerging Chinese culture. But although its appearance changed, it retained its fundamental reptilian body form and essential link to water as a reminder of its evolutionary past.

A Dragon's Life

Ecology and Behavior of the Chinese Alligator

On 21 July 1999 we were driving in a two-car convoy along back roads from the alligator breeding center in Xuancheng. Our destination was Zhuang-tou, where in 1997 we had seen the remains of two nests whose eggs had been collected and taken to the breeding center. While anathema to the survival of the wild population, the collection of eggs from the few wild nests had been routine since the breeding center was established in 1979. Farmers knew where alligators nested, and notifying the breeding center of the presence of a nest would bring them a much needed cash reward. The breeding center paid for eggs from wild nests because they believed that, if they hatched, there was no hope for the survival of the alligators that would be produced. And therein lies the rub. The act of removing wild alligators and bringing them into the breeding center was symptomatic of a failure to provide a future for the surviving groups of alligators. The wildlife authorities in China had given up on them.

Zhuangtou was a special place; in the mid- to late 1990s it was the only site where more than one nest had been produced in any one year. It was the only location where more than one adult female was known to live. Yet despite this remarkable fact, it was not, at that time, among the 13 sites designated for the protection of the Chinese alligator. (It was included in the revised listed of designated sites in 2006.) As in many of the designated areas, alligators at Zhuangtou lived in an assemblage of small ponds set among rice paddies, adjacent to a series of low hills, cloaked in pine plantations. It was into these hills that the alligators climbed to make their nests.

The date 21 July was an auspicious one for us. Exactly one year before we had been involved in a car accident that brought our 1998 field survey to an abrupt end. We had now returned to take up where we had left off, but Anhui forestry bureau reports from the previous year, a very wet one, were alarming. Not one wild nest had been reported. Perhaps for the first time in the entire evolutionary history of the species there had been no reproduction in the wild population. It was an ominous start to our work.

We turned off onto a side road. Local farmers were drying their newly harvested rice on the concrete edge of the road as we continued up to a small village at the base of a chain of hills. The air was heavy with the haze, a product of the heat, the humidity, and the smoke from burning rice fields that seems to cling to this landscape for weeks on end. Our AFB counterpart, Yueyang Long, went in search of the farmer who tended the fields closest to where the alligators were found and who had guided us to the nests in 1997. His wife was not happy to have him set off with us, yelling that he had to walk their buffalo out to graze. Nevertheless he accompanied us after we negotiated suitable compensation. We hiked up through valleys filled with the lush geometrical green of rice paddies, navigating a labyrinth of narrow trails that formed the margins of the fields. The farmer told us that he had been busy and had not had much time to look for nests that year, but that as of yet none had been seen by anyone he knew. We searched around one spot where two years before we had seen the remains of a nest whose eggs had been collected by ARCCAR staff, but found no traces of nest or alligator. The second nest site in 1997 was higher up on a ridge above two of the highest ponds, which like many ponds used by alligators had been created to store water for seasonal flooding of the rice fields in the valley below. The nest we had seen at this site in 1997 was unusual even by the unconventional standards of Chinese alligator nests. The small ponds at Zhuangtou looked marginal, at best, as the home for an alligator. The bare, clayey soils and nearly transparent water supported little in the way of aquatic vegetation, or seemingly of potential prey for alligators. Yet walking up a small defile we entered a narrow but heavily vegetated gap that contained the unmistakable signs of an alligator trail. At the head of this path we found a small burrow dug into the hillside, from which trickled a meager stream of water. One of the 1997 nests had been found a short but steep climb up from this burrow, at the base of a pine tree. We found what remained of that old nest mound, but there was no recent activity. Broadening our search, we fanned out over the hillside, amid dense bracken ferns and a ground cloaked with pine needles. Much to our delight we soon found the distinctive signs of a nest, a small area at the base of a tree whose ground litter had been scraped up into a small mound of pine needles. This was the first nest that our team had ever found, and soon the two students from East China Normal University, Lijun He and Youzhong Ding, were helping us inspect it. Carefully opening the nest mound from the top, inside we found 18 ivory colored eggs, some of which were being tinted a brown color by the moist pine needles around them. Under the watchful eye of the farmer and our police escort, we counted and measured a sample of the eggs, each delicately marked with a pencil so we would not change their vertical orientation, which can kill the embryo. Before leaving, we implanted a small temperature data-logger.

A few days later we were back at the breeding center and discussed with the head of the AFB the need to leave wild nests in place. How could the wild population ever maintain itself, much less recover, if eggs were being routinely extracted from the wild population? They agreed not to collect this nest, as well as another we subsequently found at Shaungken. They did eventually collect the Shaungken eggs after a local farmer opened the nest to steal our data-logger. The Zhuangtou nest was left, but failed to hatch, a result of lethally low incubation temperatures, likely the result of the nest being made in such an exposed location and built with pine needles. Despite our best efforts to add to the number of wild alligators at this site, the alligators had seemingly been pushed beyond their biological limits.

In Fauvel's 1879 description of the Chinese alligator there were detailed explanations of the species' history and discovery, as well as an erudite report on its morphology, but at the time next to nothing was known about its ecology or behavior. By the early twentieth century, the alligator's range and numbers had been so reduced that opportunities for studying behavior and ecology were limited. The first biological investigations of Chinese alligators were on the seasonally flooded grassy plains along the lower Qing Yi and Shui Yang rivers that drain southeastern Anhui Province, flowing into the Yangtze River near the city of Wuhu. This region, and the hills that these rivers drained to the south, was the last area where Chinese alligators could be found in any numbers in the twentieth century. While this low-lying area flooded extensively during the summer, the water levels retreated dramatically in the winter and the marsh vegetation was cut or burned, revealing the locations of burrows used by alligators for hibernation. Most of the information we have on wild Chinese alligators during this period comes from a succession of winter visits to this area by American and Chinese scientists who paid local farmers to dig alligators from their dens (Chapter 3). While the snippets of information that emerged from these visits offer a tantalizing glimpse of the alligator's ecology, they are often difficult to interpret as they were not carried out with the same rigor that we expect from modern wildlife investigations. Subsequent studies, including ours, have been limited mostly to breeding centers or small groups of individuals living in highly altered habitats. What we can learn about alligators from these areas may not be directly applicable to understanding the biology of alligators in the past, when habitat and alligators were abundant, nor may it offer the best guidelines to plan restoration efforts.

In this chapter we piece together information on the ecology of Chinese alligators from several sources. These include observations made by scientists in the early and mid-twentieth century, as well as the considerable body of literature published in Chinese, largely from the staff and researchers at the Anhui Research Center for Chinese Alligator Reproduction, the largest breeding center for this species. Observations conducted over the last 20 years outside the breeding centers have been limited largely to opportunistic reports of the last remaining wild alligators. To this body of information we have added the results of our own work in Anhui Province since 1997 on wild alligators.

This still leaves large gaps in the collective knowledge about these animals. To fill these gaps, or at least shed some light on other important aspects of the biology of the Chinese alligator, we turn to its closest living relative, the American alligator. While the American alligator is considerably larger than *A. sinensis*, these two species share a close evolutionary relationship (Chapter 3), as well as ecological and behavioral characteristics that derive from their being the only two crocodilians found in areas with temperate climates. To this end we use studies on the American alligator as a point of reference for extrapolation and discussion concerning the ecology of the lesser-known Chinese species.

At the end of this chapter we synthesize our knowledge of ecology and behavior in a reconstructed account of a year in the life of the Chinese alligator in the seasonally flooded grasslands near Wuhu, the last area where it was known to occupy relatively natural habitats. This account pulls together what we know, or can surmise, about the species' annual cycle of activities, dictated as they are by the seasons of its temperate landscape.

Habitat

The question of habitat is critical to understanding the ecology and behavior of the Chinese alligator. Here we are faced with a major dilemma—we do not have a clear idea of what actually was the natural habitat of the Chinese alligator. We know relatively little about the former habitats occupied by Chinese alligators as they were already close to extinction by the time biological studies were initiated. In the southeastern United States, American alligators tend to be creatures of marshes and the margins of lakes, but they are quite adaptable and can be found in a variety of habitats. Based on what we know about American alligators, we can surmise that A. *sinensis* also used a diversity of wetland habitats, likely centered on shallow water wetlands in the floodplain of the Yangtze River and its tributaries in Eastern China.

It is difficult to precisely define the ecological characteristics of American alligators because they are found in such a wide variety of habitats (Box 5.1). They are principally found in freshwater habitats, but in some areas are common in coastal brackish water habitats and have occasionally even been seen swimming offshore in the ocean (Elsey 2005). Living in such a diverse assemblage of habitats results in significant ecological diversity between populations in different parts of the species range. For instance, in the northern part of its range the alligator remains inactive for several months each year as a result of low ambient winter temperatures. On the southern tip of Florida, instead of the annual fluctuations in temperature it is the tropical pattern of dry and wet seasons that shapes annual activity patterns. Differences in habitat type can also result in dissimilar potential prey items and changes in the diet of the American alligator throughout its range (Delany et al. 1999), as well as considerable variation in important aspects of their reproductive ecology.

Life in a Hydrologically Variable Environment

The Yangtze River basin is dominated by the eastern Asian monsoon, which produces a highly seasonal pattern of rainfall, much more so than that in the southeastern United States. The monsoon is an Asian climate pattern that refers not only to the annual rains during the wet season but also to the seasonal alternation between wet and dry periods that is caused by the juxtaposition of the world's largest land and water masses, along with the presence of the world's greatest mountain system. Cold, dry winters are a product of a high pressure zone and the frigid continental air masses that dominate central Asia. Conversely, during the summer, the enormous Asian landmass heats more rapidly than the oceans, creating a prevailing low pressure system that draws in moisture-laden air from the Pacific. As the seasons shift from winter to summer, the cold and dry northwesterly winds are replaced by warm and wet air from the southeast and the predictable rains that made East Central China such a suitable place for agriculture. The Himalayas wring even more water from the summer monsoon as the air rises and cools, falling as rain at lower altitudes and snow higher up. This water adds to the rains in Eastern China and swells the Yangtze, which absorbs this extra volume by overflowing into low-lying buffer areas, the river's floodplain. We can see the effects of these highly seasonal rains, and how they differ from those experienced by American alligators in the United States, by comparing average monthly levels of streamflow in the major rivers, the Yangtze and

BOX 5.1

Habitat Variability among American Alligators

When people think of American alligators it is frequently the Everglades, a vast subtropical wetland at the southern tip of the state of Florida, that comes to mind. While it is true that there are a lot of alligators in the Everglades, it is actually a fairly harsh habitat for them (a too-warm subtropical climate with generally very low levels of aquatic productivity), and alligators in the Everglades tend to grow much slower and have smaller clutch sizes than in other parts of their range (Jacobsen and Kushlan 1989; Dalrymple 1996). North of the Everglades, alligators inhabit a great diversity of wetlands throughout the coastal plain of the southeastern United States, ranging from isolated wetlands in south Texas to streams and swamps in southeastern Oklahoma and central Arkansas and the coastal wetlands of North Carolina. Within this area, American alligators are typically found in freshwater marshes and lakes with an abundance of herbaceous vegetation. Here alligators find suitable cover, prey, and places to nest. In areas where water levels fluctuate substantially throughout the year, alligators create "gator holes," or depressions, that retain water throughout the year (Box 5.3).

American alligators, which are extremely adaptable, are commonly found in freshwater and even brackish water wetlands. They can often be found in rivers and large streams (but not usually in small streams), but typically at lower densities than in marshes or lakes. In these fluvial habitats they are frequently, but not always, found in slow-moving sections or in parts of rivers associated with floodplain lakes or fringing marshes and hardwood or cypress swamps. Alligators also use man-made water bodies such as reservoirs, canals, abandoned rice fields, coastal mangrove swamps, and other brackish water habitats, including salt marshes where they have been found nesting in areas with salinities up to nearly full strength sea water (Wilkinson 1984). The highest densities of alligators in South Carolina and Louisiana have been reported to be in intermediate salinity marshes along the coast (Wilkinson 1984; McNease and Joanen 1979).

The adaptable nature of alligators and growing populations of alligators and people have combined to bring these two species into conflict. Today it is not unusual to find alligators in canals or ponds next to homes, even in ponds on golf courses and in some suburban area. Most states have implemented nuisance alligator programs to regulate the removal of alligators from populated areas.

It is likely that Chinese alligators were once found in a diverse array of wetland types throughout their historical range along the lower Yangtze. This would have resulted in regional differences in their ecology and behavior. But as a result of conversion into agricultural lands and the extensive development of flood control structures, this habitat diversity no longer exists. The one area where historical populations were known to be, in the vicinity of Wuhu in Anhui Province, was a seasonally flooded marshland in an area where two rivers emptied into the Yangtze. In the 1920s Clifford Pope, collecting specimens for the American Museum of Natural History, was able to obtain specimens excavated from burrows in a grassy plain by the side of the Chingshui Ho, a river that flows from the south through Wuhu and into the Yangtze (from Schmidt 1927). Pope's description of the plain noted that it was "treeless, and even the grass is sparse. At this time of the year [winter] the river flows between steep banks, perhaps 25 feet high."

During a previous summertime visit to the same location Pope had found the plain completely flooded and was told by local farmers that it was very difficult to locate alligators. Hsiao (1934) visited the same area in the early 1930s (also during the winter) and reported that the area had many dried creeks and ponds. Alligator burrows were found around small ponds that dotted the floodplain but not along the riverbanks. By the time Chu-chien Huang began his work in the 1950s, he characterized the habitat of the alligator as "marshes, sandy land, and porous dry riverbeds of the hilly country in southern Anhui." He noted that in these areas they were not subject to the annual floods of the larger rivers (C. Huang 1982).

The lower Yangtze region, however, was historically characterized by an extensive floodplain where alligators were likely also found in a diverse assortment of ponds and lakes, and possibly in some riverine habitats. The occurrence of Chinese alligators in coastal habitats has been poorly documented; these were likely some of the first areas from which they were extirpated. Nevertheless, there is archaeological evidence that alligators lived in coastal regions around Hangzhou Bay (Chapter 6). Like American alligators, Chinese alligators were once widely distributed, and we assume they lived in a variety of ecological settings that today we can only guess at. In trying to reconstruct the ecological characteristics of the Chinese alligator, two things are important to bear in mind: It is living close to the northern extreme of its range in terms of temperature tolerance, and its habitats are characterized by a greater seasonal fluctuation in water levels than that of the American alligator.

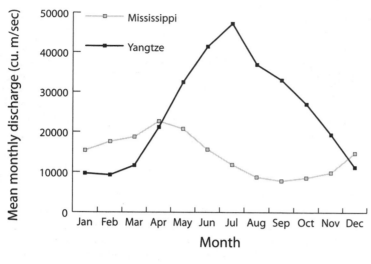

Figure 5.1. Mean monthly discharge of the Yangtze (at Datong) and the Mississippi (Vicksburg). Data from Vörösmarty et al. 1996.

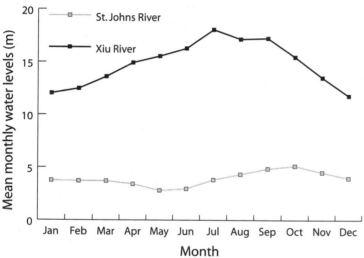

Figure 5.2. Mean monthly river level of Florida's St. Johns River (near Melbourne) (1940–2005) and the Xiu River in Jangxi Province, China (1987–1991). In an average year the St. Johns fluctuates 2.3 m; the Xiu, 6.3 m.

the Mississippi rivers (Figure 5.1) and in smaller river systems such as the St. Johns in Florida and the Xiu, a Yangtze tributary that empties into Poyang Lake (Figure 5.2). Along the Yangtze, this annual pattern of precipitation and flooding obliges everything from plants to people to adapt to the summer inundation of the extensive alluvial plain. While American alligators inhabit a variety of seasonally fluctuating wetlands, nowhere do water levels vary as much on an annual basis as they do along the Yangtze. Accordingly, one of the greatest differences in the habitats used by American and those that historically would have been used by Chinese alligators stems from the seasonal pattern of flooding along the Yangtze River.

How do crocodilians adjust to such extreme annual fluctuations in water level? Variation in water levels affect many aspects of the ecology of alligators, and the key to understanding the response depends very much on the species' reproductive strategy, specifically how that species makes its nests. Crocodilians use two basic types of nests, and the use of one or the other type of nest fundamentally shapes many aspects of nesting ecology, including the timing of nesting with respect to seasonal changes in water levels. Some of the true crocodiles, as well as the gharial, deposit their eggs in holes excavated in riparian substrates, usually in carefully selected areas of suitable soil along riverbanks or in sandbars. However, most crocodilians, including both species of alligator, lay their eggs in "compost" mounds of vegetation and other organic matter constructed by the nesting females. For both of these nesting strategies the annual timing of reproduction is usually tightly related to environmental factors, with rainfall and water levels being among of the most important. Most hole-nesting crocodilians, including the Nile, Orinoco, and mugger crocodiles, live in seasonal rivers and nest during the annual low water period when alluvial beaches are exposed. Mound-nesting crocodilians, on the other hand, construct nests in a variety of habitats, ranging from lakeshores to marshes, but select microhabitats that are unlikely to flood. Because they are more adaptable in the types of areas where they can nest, mound-nesting species also show greater flexibility in seasonal timing of nesting, and this may vary between species as well as between populations of the same species. For instance, South American caiman living in areas of highly variable water levels have different seasonal patterns of nesting. In the hyperseasonal savannas of South America, yacare caiman in the Pantanal (*Caiman yacare*) and spectacled caiman (*C. crocodilus*) in the llanos nest during the height of the rainy season when environmental flooding is at its peak (Crawshaw 1987; Cintra 1988; Thorbjarnarson 1994). In these areas sufficient vegetation for the construction of nest mounds is only available during the wet season, and nest mounds are made on elevated microlocations to reduce the potential for nest flooding. Nesting during this time of year also makes it more difficult for egg predators to find the nests. Conversely, in the western Brazilian Amazon, both black (*Melanosuchus niger*) and spectacled caiman live in flooded forest habitats where water levels fluctuate up to 15 m on an annual basis (Da Silveira and Thorbjarnarson 1999; Thorbjarnarson and Da Silveira 2000). In many of the *varzea* habitats there is little to no dry land when water levels are high, so both species nest during the annual low water period, but use different strategies. The smaller spectacled caiman will move inland, sometimes over 1 km, seeking high ground that is unlikely to flood before the end of egg incubation. The adult female black caiman, however, occupies perched floodplain lakes where water levels are quite stable for much of the dry season (Thorbjarnarson and Da Silveira 2000).

Both the American and Chinese alligators are more temperate than tropical in their habitats, and reproduction is constrained not so much by water levels as by temperature (Lance 2003), with a fairly tight window of opportunity for successful nesting during the year. American alligators nest in early summer when air temperatures are relatively high and sufficient herbaceous vegetation is available for building nest mounds. Incubation takes approximately two months. Hatchlings emerge in late summer just as ambient temperatures begin to drop and have a relatively short activity period before they must hibernate for the winter. Chinese alligators face the same temporal and temperature constraints on nesting. In addition they probably had to deal with a much more hydrologically variable environment than their American congener, and it is likely this played a major role in determining where alligators nested, which in turn would have determined the types of habitats selected by adult females. The nesting ecology of Chinese alligators may then have been a "hybrid" of those seen in American alligators (temperature constraints), and the tropical caiman (significant seasonal fluctuations in water levels).

Today, the volume of water in the Yangtze River increases fivefold from the winter low water period to the summer floods (Figure 5.1). The local terrain of floodplains would have dictated patterns of seasonal flooding, and played a determining role in where alligators could nest. Water levels in the main river channels and areas immediately adjacent would experience much greater annual variability in water levels than in more elevated seasonally flooded lake and marsh habitats. Water levels in these latter habitats would have been more stable, and these wetlands, perched on the floodplains well above the rivers would offer more suitable nesting habitats. Areas like the Wuhu marshes described by Pope and others in the early twentieth century were extensively flooded in the summer and virtual deserts in the winter, much like the hyperseasonal savannas where caiman nest in South America. The rivers and streams were rising and falling 7.6 to 9.1 m annually. But many of the marshes and ponds were in isolated depressions that would only fill with rain or with floodwaters when rivers were at their highest and would hold their water while the floodplain rivers and streams were in constant flux. In these perched marshes and ponds, water levels would have remained much more stable (with respect to the river level) during the warm, summer nesting season. Similar to American alligators and caiman, successful nesting of Chinese alligators in these floodplain habitats would likely have depended on finding small, raised microhabitats next to these hydrologically perched ponds and marshes. These may have been the types of sites where Pope and others were finding their "colonies" of burrows during the winter, when the landscape would have looked very different. Another, perhaps fortuitous, environmental factor is that, at least along the main branch of the Yangtze and some of its major tributaries, the mid-July nesting period, which is largely defined by the annual temperature regime, coincides with peak water levels (Figures 5.1 and 5.2). So at least in some parts of the floodplain, a female alligator building a nest on a dry (unflooded) site in July is selecting an area that is unlikely to be inundated later in the year.

Life in a Temperate Climate

In the United States, American alligators are found across a broad range of latitudes from 36° N in coastal North Carolina to 16° N in southern Florida, and throughout this range seasonal temperature variations similarly define the annual activity cycle of alligators. At the southern end of this distribution it rarely gets cold enough for alligators to need a period of winter inactivity, and the annual activity cycle is governed principally by subtropical wet and dry cycles. At the northern limit of its range the annual activity period of American alligators is reduced to about eight months by cold winter temperatures. Like the American alligator near its northern limits, Chinese alligators' activity is reduced for much of the year due to low temperatures. As air temperature falls in October, alligators stop feeding, spend more of the day basking and more of the night hidden in burrows. By early December alligators only emerge from burrows on relatively warm, sunny days, and by mid-December they have usually entered a deep hibernation and torpor, rarely moving. Alligators emerge from this deep sleep in late February and start to leave their burrows on warm days in March, but are not very active. As air temperatures begin to increase in the spring, Chinese alligators become increasingly active. As described by Hsiao (1934):

> The process of entering into hibernation is a very gradual one. As the weather gets colder in the fall the alligator stays more and more in its hiding place. It is at noon time and in the early afternoon that it comes out to bask in the sun, but it goes into its cave again before or soon after sunset. It does not come out at all when the weather is cloudy or rainy. Basking becomes less and less frequent as autumn gives place to winter and hibernation then becomes complete. In the spring it comes from its lethargy in a very similar manner. Once in a while it comes out to bask at noon or early afternoon and returns to its bed chamber at night. As the weather becomes warmer it stays longer and longer out. Sometimes it even stays away from its subterranean home overnight by submerging itself in water completely.
>
> This habit of basking in the sun both before and after hibernation is directly opposite to its nocturnal habit. During the warmer seasons it is always nocturnal. It hides either in its cave or underwater for the greater part of the day, and even when it appears at the surface of the water, it immediately sinks back into the water without a sound and with scarcely any ripples at the sign of anything suspicious and stays underwater for hours, outwaiting the patience of any observer.

In the spring, alligators are usually first seen out of their burrows in March to April when the temperature at the upper levels of the burrow begins to exceed that deep within the burrow (C. Huang 1982). After their spring emergence alligators are largely diurnal, basking on sunny days and avoiding activity during the cooler nights. As temperatures rise in May and June they become increasingly nocturnal (C. Huang 1982; Watanabe 1981). Courtship, bellowing, and mating start in early June and peak in mid-June. Females nest in late June and early July, and eggs hatch in September. Hatchlings have at most two months of activity before hibernating in burrows with their mothers.

Alligators live in colder climates than any other crocodilian, and they exhibit a number of behavioral, and perhaps physiological, adaptations that allow them to do so. While alligators can tolerate colder temperatures than other crocodilians (Brandt and Mazzotti 1990; Brisbin et al. 1982), they cannot survive body temperatures at or below freezing. Minimum winter temperatures may limit the northern distribution of alligators via direct mortality. An alternative explanation is that seasonally cool temperatures in the fall may constrain the period during which nesting and egg incubation can be successfully accomplished (Hagan et al. 1983). While young alligators are presumably more vulnerable to cold-related mortality (because they have smaller body mass and cool faster) even adult alligators have been killed by low temperatures. During one particularly cold period in the winter of 1779 to 1780, a Captain William Phelps found a number of dead alligators along the shores of the Big Black River in what is today the state of Alabama (Phelps 1804). A sudden drop in temperature from a massive cold front in coastal Texas in December 1924 resulted in the mass mortality of American alligators, including adults (Weigelt 1989). An account from Arkansas attributed the death of a large male American alligator (3.12 m TL) to cold weather (Trauth and McCallum 2001). Nevertheless, American alligators show a remarkable ability to survive cold periods during the winter. Adult American alligators can tolerate body temperatures as low as 5.0° to 0.4°C (Brandt and Mazzotti 1990; Brisbin et al. 1982). These are lower body temperatures than can be tolerated by more tropical species (e.g., *Caiman crocodilus*; Brandt and Mazzotti 1990). This suggests that alligators have physiological mechanisms that allow them to withstand low body temperatures that are lethal to other crocodilians. Alligators also show behavioral adaptations for dealing with cold ambient temperatures. Tropical caiman when presented with cold ambient temperatures will leave the water and attempt to move overland, which can quickly result in lethal body temperatures. Alligators, on the other hand, remain in the more thermally stable environment of water (Brandt and Mazzotti 1990).

The lowest annual temperatures in the Northern Hemisphere are generally in the month of January, and the northernmost extent of the American alligator's distribution corresponds fairly well with the January mean temperature isotherm of 7.2°C. But if direct mortality during the winter is what most limits the alligator's northern range, then minimum January temperatures, not average temperature, are more likely to determine the northern edge of the range of the species. In fact the northern limits of American alligators correspond well with the minimum January temperature isotherm of –9.4°C (Neill 1971), as well as a mean minimum January isotherm of –1°C. Along the Atlantic coastal plain this is consistent with a northernmost distribution of the American alligator in northeastern North Carolina or southeastern-most Virginia. Historical records suggest that alligators may at one time have been found in the Dismal Swamp (Neill 1971) in the southeastern extreme of Virginia at latitude 36.5° N. Alligators were never known to be abundant in this area, which has been significantly impacted by human development. No alligators have been reported from the Dismal Swamp in recent years, but since 2000, alligators have extended their range to an area north of the Albemarle Sound in North Carolina (J. Groves, pers. comm.). Inland, in the Mississippi River drainage, American alligators reach their northern limits in Arkansas, where some individuals have been found up to latitude 35.8° N (Trauth and McCallum 2001). Isolated cases of American alligators surviving for periods of time even farther north have been reported. There is a record of a juvenile American alligator (ca. 50 cm TL) that survived over six winters in Castle Shannon, Pennsylvania (lat 40°22′ N), far to the north of the current distribution of the species. It grew to 1.25 m before it was shot (Barton 1955). Mean air temperature during the coldest months at this site was –5.5°C. The ability of individual alligators to survive for extended periods in these areas suggests that the northern limits of the distribution of American alligators may be limited by factors other than direct mortality. Another potential factor is how temperature may limit the seasonal window of opportunity for nesting (Klause 1983), with falling environmental temperatures during the end of the incubation period resulting in nest failure.

The Chinese alligator does not range nearly as far north as American alligators. The surviving groups are found at about latitude 30.5° N, which corresponds roughly to the Florida-Georgia boundary in the United States. But because of climatic differences between Eastern China and the southeastern United States, Chinese alligators are probably at or near the northernmost limit of their distribution. That is, American alligators can live farther north because winters on the eastern seaboard of the United States are warmer than those of Eastern China at equivalent latitudes

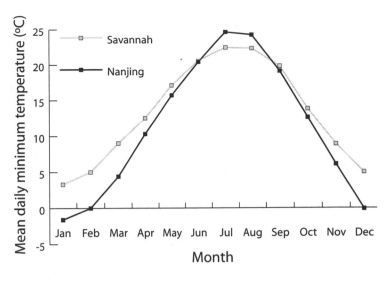

Figure 5.3. Mean minimum daily temperatures by month for Nanjing and Savannah, Georgia. Both are located at latitude 32° N.

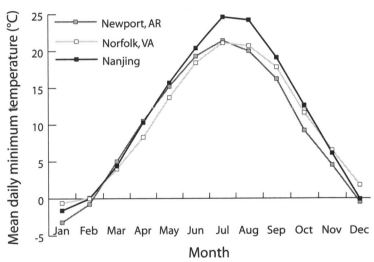

Figure 5.4. Mean daily minimum temperatures by month for Nanjing, Jiangsu Province, and two cities in the eastern United States at the northern edge of the distribution of American alligators. Data represent values for more than 30 years. From www.world climate.com.

(Figure 5.3). For instance, in recent historic times the northernmost limit of Chinese alligators was approximately equivalent to that of the city of Nanjing (lat 32.0° N). However, at a corresponding latitude in the southeastern United States, the mean annual minimum daily temperature is 2.0°C warmer (Savannah, Ga., lat 32.1° N), principally due to warmer winter temperatures. To find areas with minimum winter temperatures equivalent to Nanjing we would have to move almost 5° farther north along the U.S. eastern seaboard, near the area of the northern limit of American alligators (Norfolk, Va., lat 36.9° N) (Figure 5.4). The environmental temperatures experienced by Chinese alligators today in southern Anhui Province (Wuhu lat 31.3° N; mean annual temperature 16.0°C) are equivalent to that of central North Carolina (Fayetteville lat 35.2° N, mean annual temperature 16.2°C), suggesting the Chinese alligators are at or near the northern edge of their temperature limits.

This raises an interesting question. If Chinese alligators are right up against their northern climate limits, why aren't they found farther south, in warmer climates, as American alligators are? The answer is a combination of physical and human geography. The area to the south of the current distribution of Chinese alligators in Anhui Province is mountainous and does not offer wetlands suitable for alligators. The coastal plain to the south of the Yangtze River delta had suitable habitat around Hangzhou Bay and possibly along the lower Qiantang River. Remains of alligators have been found in this area from excavated Neolithic settlements, but alligators were apparently extirpated here almost 1,000 years ago, likely as a result of human population pressures (Chapter 6). Farther south, along what is today Zhejiang Province, the coastal plain is extremely narrow and while Chinese alligators may once have been found there, the combination of human population pressures and limited habitat would have made them very susceptible to extirpation. Farther inland from Anhui Province, the historic distribution of the Chinese alligator followed the Yangtze River, which flows to the southwest. In past times alligators were found in the alluvial habitats around the Dongting Lake in northern Hunan Province, situated at about latitude 28° N, some 300 km farther south than today's groups of alligators in Anhui Province. This area would have offered suitable wetlands habitats and moderate temperatures (Changsha, lat 28.3° N, mean annual temperature 17.3°C; www .worldclimate.com). Nevertheless, alligators disappeared from there centuries ago, presumably as a result of human population pressures and habitat loss. The conservation implications of these temperature limits on the distribution of Chinese alligators are discussed in Chapter 6.

American and Chinese alligators survive periods of low temperature by using burrows as winter refuges. These underground retreats provide thermal buffering from aboveground temperature extremes, and the use of these burrows is an extremely important component of the ecology of Chinese alligators (see below for further discussion). But burrows are not the only behavioral strategy employed by alligators during cold periods. In areas where water levels are relatively stable or where the habitat prohibits digging burrows, American alligators are believed to pass periods of cold weather in deep water or under vegetation. When exposed to dangerously low environmental temperatures, American alligators are known to use a typical "submerged breathing" posture (Chabreck 1966; Smith 1979; Brisbin et al. 1982; Hagan et al. 1983) by remaining submerged in shallow water or at the entrance to a burrow and periodically raising the head to breath. This behavior, which allows alligators to survive short periods of subfreezing temperature, has been observed in bodies of water that have iced over, with ice-free breathing holes maintained by the alligator periodically raising the tip of its snout to the surface. Observations on both species of alligators held together in outdoor pens in northern Florida suggest that the Chinese alligator is at least as cold tolerant as the American species (D. Kledzek, pers. comm.), and a similar "icing" behavior has been seen in Chinese alligators in breeding centers in China (Z. Wang, pers. comm.). American alligators less than 50 cm TL apparently cannot keep breathing holes open in ice and are, therefore, more vulnerable to mortality as a result of sudden cold snaps (Brandt and Mazzotti 1990). Smaller alligators also have less thermal inertia than larger individuals and would more rapidly reach lethally cold body temperatures.

BOX 5.2

Thermoregulation

The regulation of body temperature increases the fitness of animals by ensuring that biochemical and physiological processes proceed at an optimal rate. Alligators are ectotherms; instead of staying warm by having a high metabolic rate, as mammals and birds do, they regulate body temperature primarily by behavioral means. This includes avoiding temperature extremes and adjusting body temperature by moving between sources and sinks of heat. Alligators climb up on shore and expose their bodies to the direct rays of the sun to warm themselves; if they become too hot they slide back into the water to cool off. In some circumstances water temperatures exceed air temperature and this can be used to avoid lethally cold winter conditions. Most reptiles try to maintain their body temperature within a preferred range. In the tropics some crocodilians will move to shore during the late afternoon to cool their body when water temperatures exceed preferred levels. Under some heat-stressed conditions tropical crocodilians will bury themselves in the mud at the bottom of shallow bodies of water to stay cool. While little is known of the preferred body temperatures of the Chinese alligator, it is probably similar to the American alligator, both being more temperate than tropical.

Classic studies of behavioral thermoregulation in reptiles were conducted mostly with small-bodied lizards and snakes, which because they have a small body mass can fine tune their body temperature by shuttling between areas where they warm up and areas where they cool off. But crocodilians (adults at least) are much larger, and their body temperature changes at a much slower rate (Lang 1987). Some relatively small crocodilians can follow a typical reptile thermoregulatory pattern by maintaining a high and fairly stable body temperature during sunny days (*C. johnsoni*; Grigg and Seebacher 2001). Although American alligators avoid temperature extremes, their body temperatures tend to oscillate regularly over a 24-hour period as environmental temperatures vary (Seebacher et al. 2003), which suggests that keeping a very stable body temperature is not necessary. Nevertheless, broad behavioral patterns to warm themselves during cool weather and cool off when it is hot are clearly evident. During the winter alligators spend more time basking in the sun to elevate body temperatures; likewise, in the summer they spend more time in the water and become nocturnal.

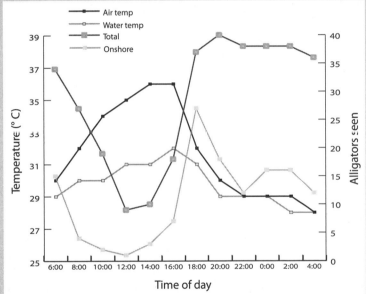

Daily summer pattern of air and water temperatures and the number of alligators in the water and on land basking at the Changxing breeding center. From X. Wang et al. 2006.

In coastal Louisiana average daily body temperatures vary seasonally from 30°C in summer to 16°C in winter. This suggests that alligators acclimatize by changing the thermal sensitivity of physiological and biochemical processes, something that has been found in other ectotherms (Seebacher et al. 2003). When environmental temperatures are high, Chinese alligators become largely nocturnal and spend most of the day in the water. Behavioral observations on captive Chinese alligators during the hottest part of the year in China (June–September) found a predictable pattern of alligator behavior: on land in the early morning hours and again in the evening (X. Wang et al. 2006) when air and water temperatures are about the same. During the heat of the day most alligators remained in the water.

Alligator Burrows

Both Chinese and American alligators dig burrows in the margins of lakes, ponds, and marshes that are used as daily retreat sites and refuges that allow them to survive cold winter air temperatures that can dip well below freezing. Considerable confusion in the literature is found with regard to the terminology used for different kinds of excavations made by American alligators. Aside from the tunnel-like burrows, alligators are also well known for their role in the creation or deepening of natural holes in wetlands, particularly in areas where there is considerable seasonal variability in water levels. These "holes" are in effect small ponds that the alligators build for themselves, holding water in what might otherwise be a dry landscape. Both the vertically excavated holes and the horizontally excavated burrows have been referred to as "dens" in the alligator literature. Neill (1971) gives an excellent overview of the types of excavations made by alligators but fails to provide an adequate terminology of these structures, instead using the term "den" to refer to "any excavation made by the alligator for its own occupancy, regardless of the excavation's shape or the season during which it is occupied." We suggest the following terminology to distinguish between the types of shelters made by alligators:

> *Hole.* A pool of water, used by alligators, whose bottom is below that of the surrounding landscape and that the alligator has played a role in creating, usually by deepening natural depressions.
> *Burrow.* Horizontal tunnels dug by alligators into raised terrain in or near wetlands.
> *Den.* An area frequented by alligators; it may have a hole, a burrow, or both.

"Gator" holes are widely recognized for their ecological importance in providing aquatic habitat for a variety of wetland fauna during annual periods of low water (Box 5.3) in the Florida Everglades, the Okefenokee Swamp, and other seasonal wetlands in the United States (Palmer and Mazzotti 2004; Campbell and Mazzotti 2004). These holes (it is not clear what percentage, if any, have associated burrows) provide important foraging habitat for mammals and wading bids (Frederick and Spaulding 1994; Hoffman et al. 1994) and also increase the spatial heterogeneity of marsh landscapes, altering plant community structure on a local scale.

From the brief historical descriptions of the habitats used by Chinese alligators in the vicinity of Wuhu, and the fact that water levels in the lower Yangtze basin varied considerably on an annual basis, it is quite likely that in the past Chinese alligators would make holes as a way to ensure water availability during periods of low water. This feature of their ecology would have been most evident on the extensive floodplain habitats from which they were extirpated before any meaningful scientific investigation got under way. Nevertheless Chinese alligators are prolific diggers of burrows, and that is what we focus on here. Many crocodilians, including alligators, as well as the mugger, the Orinoco, the Nile, the American and Morelet's crocodile, dig tunnels to create subterranean refuges (Pooley 1969; Gupta and Bhardwaj 1995; Platt 2000). Aside from their use by alligators as shelter during period of cold weather, as also used by alligators and other species to aestivate during dry periods, as diurnal or nocturnal refuges when inactive, and to hide from predators or aggressive conspecifics (Platt 2000). Both adult and juvenile crocodilians dig burrows (Pooley 1969), and adult females frequently have burrows near their nests, which they use as shelters for neonates. Burrows can be quite long (10–15 m or more) and some have terminal roomy chambers that allow the animals to turn around (Guggisberg 1972). Crocodilians may occupy burrows over extended periods of time. Toward the northern edge of their distribution, American alligators overwinter in burrows for periods of up to 5 months each year (Neill 1971). In the Sahel region of Mauritania, Nile crocodiles spend most of the annual dry season (up to 10 months a year) in burrows (T. Shine et al. 2001).

One of the most detailed accounts of the construction and use of holes and burrows was made by E. A. McIlhenny (1935) based on his experiences with American alligators in the seasonally flooded Louisiana coastal marshes. "In such locations alligators dig considerable holes, sometimes four to six feet in depth and with surface openings ten or more feet across. These holes are connected with underground dens which they have dug out, sometimes as much as forty feet long; into which they retire on the approach of danger, and in which they usually spend the cold winter months. These dens are usually kept filled with water by the summer rains after the spring freshet recedes, or are dug to the water table which is never more than a few feet below the surface in the swamps and marshes."

Based on his lifetime of experience in the Louisiana marshes, McIlhenny provided a number of insights concerning the construction and use of these burrows. He noted that they are usually located under the banks of streams or lakes, but that at times he found them in open marsh, several miles from sources of permanent water, with the occupants living apparently normal lives. McIlhenny also found that hatchling alligators would share their mother's burrow to survive their first winter, and that during a period of extreme drought alligators would move to new locations and dig dens where water was available: "I have known instances where large, old alligators who had their dens at small ponds in the marsh never [left] these ponds, relying for food on such animals, birds and reptiles as would come to the hole for water, or in search of food. Most alligators, however spend only the cold weather at these dens; leaving them for some stream or other open water as soon as Summer begins, and on the approach of cold weather going back to them for the winter."

Burrows, or holes, are important focal points in how alligators use their landscape. Radiotelemetry studies of American alligators have highlighted the importance of dens for both male and female adult alligators. Hagan (1982) found that within their home ranges adult alligators have one to three core activity areas that centered on dens. Alligators spent most of their time in these core areas and when traveling from one to another used rapid and deliberate movements. In the coastal marshes of Louisiana, females make nests near den sites and remain close to these throughout the incubation period. After the nests hatch, a female may move, with her young, from one den to another (Joanen and McNease 1989). In tidal areas of coastal South Carolina, alligator holes made by nesting females sometimes provide the only source of freshwater needed by hatchlings, as the holes are filled with rainwater rather than by tidal river flow. The construction of these holes allows nesting females to use habitats that would otherwise be unsuitable due to high salinities (Wilkinson 1984).

Observations on a number of species indicate that burrows are dug with the head and the forefeet, and by moving the body and tail in broad sweeping motions the animals can push or waft away the sediments that accumulate around the mouth of the burrow. One of the most detailed account of this behavior was given by McIl-henny for the construction of a new den by a large male American alligator that, as a result of drought, had been forced to abandon his usual site.

> Before we got to the pool we heard a considerable commotion on the water, and from the noise I surmised the alligator was deepening the pool for its winter den. Proceed-ing very quietly to a point from which the water was visible through the underbrush, I could see the back and tail of a very large alligator whose head was hidden under the bank, and whose tail was strongly sweeping from side to side. This little pond was not more than twenty-five feet across and about eighty feet long, and was completely surrounded by trees and had in it not more than ten inches of water. The alligator had not had time to be at work for more than three hours, as it had freshly arrived at the pond, and had not done any work when seen by my man earlier in the day, but when I arrived it had already cleared the mud and trash from a space about twelve feet wide by fourteen feet long, and had made it at least twelve inches deeper than the rest of the pond, sweeping the soft material and fallen branches from the bottom of the pool into considerable flats to each side of where it was working, and had torn a hole under the bank at least two feet back.... In excavating under the bank, it would tear the earth and roots loose with its mouth, and back out with great mouthfuls of this soggy material which it deposited clear of the bank in the water, and then brushed it aside with strong side waves of its tail. When large roots were encountered, they would be grasped with the jaws and the entire body jerked back until the roots were either bitten off or torn out. Such power was used in this root-breaking operation, that we could see the trees whose roots were being torn out shake to their very tops, and some of them were a foot or more through at the base. In this way roots eight or ten inches in circumference were removed. As the hole got deeper the loose material would be pushed back and out by the hind feet. I watched him at his work for about four hours, and by the end of that time only about two feet of his tail was sticking out from under the bank.

The burrows excavated by Chinese alligators appear to be more extensive and com-plex in structure than those made by any other crocodilian, and play an essential ecological role. The ability of the Chinese alligator to remain hidden in underground burrows is also undoubtedly one of the reasons why it has been able to survive in modern agricultural landscapes.

Descriptions of burrows were one of the few ecological notes made on Chinese alligators by biologists in the early 20th century because most of the specimens they obtained for museums or zoos were captured during the winter by digging out hiber-nating animals. Pope (1940) reported a "group" of burrows on a bare plain adjacent to the natural levee along a riverbank less than half a mile from a large village near Wuhu and a similar description was made by Hsiao (1935) (Chapter 3). C. Huang (1982) provides this description of Chinese alligator burrows, or dens as he refers to them.

> All dens of Chinese alligators are located close to marshes and swamps. The depth and structural form of their dens are commensurate with the required subterranean

Figure 5.5. Adult male Chinese alligator at the entrance to its burrow in Changxing, Zhejiang Province.

temperature and humidity for their hibernation in cold winter. The deepest point of their caves reaches the subterranean water level in winter. Each adult alligator has its own den, but each young [alligator] shares a den with its mother. The season for digging dens is from May to August. Chinese alligators start digging the burrow by removing the crusty topsoil with [the] claws of their front feet, and shoving the soil aside and boring a hole by forcibly moving their head forward and backward. It takes a long time [to] complete a den. The den of an old Chinese alligator is generally long and intricate. The male and female live apart, and the burrows are somewhat different in structure. The young alligator, though sharing a den with its mother, can dig a side den 2 m in length by itself.

The structure of burrows reflects their use as seasonal/diurnal refuges for alligators. Some burrows are quite long, with a complex structure (Figure 5.6), and may have up to three entrances and deep sections that retain pools of water in the winter. Chambers in the burrow are large enough to allow the alligator to turn around, so it can enter and leave headfirst. Alligators will enlarge terminal chambers in the burrows, which can be up to 150 x 70 cm in diameter; it is in these parts of the burrows where alligators generally overwinter (B. Chen 1985). Entrances can be above or below the surface of the water, depending on seasonal water level fluctuations. Most (72.5%) burrow entrances are partially or completely hidden in vegetation and 85% are within 5 m of the water's edge, with entrances having a tendency to face south, presumably to facilitate direct exposure to the sun in the fall and spring (Y. Ding et al. 2003). During our alligator surveys in Anhui Province we observed that islands in small ponds or reservoirs were a favorite location for the construction of burrows. These sites are less accessible to people, and the tendency to dig burrows along the shores of islands may simply reflect an avoidance of humans. The length of most burrows in the wild is 10 to 25 m, with an average opening width of 39 to 41 cm and height of 31 to 36 cm (B. Chen et al. 2003). Burrows of captive alligators are not as long, which may reflect environmental and soil characteristics. In the ARCCAR breeding center the mean length of 10 burrows measured by Xia and Wu (2005) was 4.15 m and ranged from 1.7 to 6.8 m long.

Burrows, which can range from 1.6 to 3 m deep, typically have an internal curving shape that is useful in blocking out light and wind (C. Huang 1982). From a sample of 17 excavated burrows Chu (1957) reported the maximum depth was 2.3 m below the surface. The depth of burrows is likely the most critical factor in determining their suitability as shelter from cold air in winter. In the wild, burrows maintain a low but relatively constant temperature of 10°C in winter (Huang 1982).

The structure of burrows used by adult male and female alligators are different (Chu 1957). Male burrows are reported to be relatively simple, with one or two openings; those of females are more complex, with several chambers in each den (Figure 5.7). One den photographed by Huang (1982) was over 20 m long (total length including all branches was over 50 m). B. Chen (1991) reported that burrow length is proportional to the age of the alligator, with older animals having longer burrows,

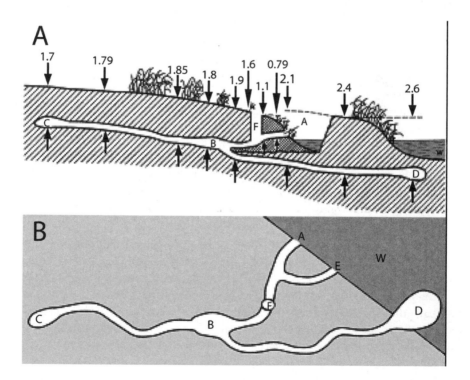

Figure 5.6. Generalized burrow structure seen in cross section (A) and from above (B). Numbers refer to vertical depth in meters from the surface to the bottom of the tunnel. A and E are entrances; C and D are terminal chambers; B is a widening of the tunnel that allows alligators to turn around; F is an air hole. The burrow is located at the edge of a pond (W). Modified from B. Chen et al. 2003.

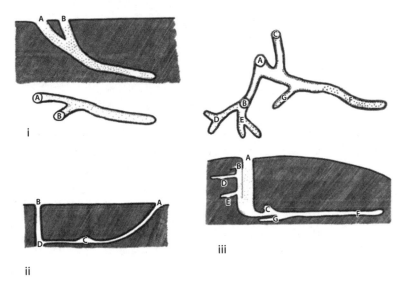

Figure 5.7. Structure of three Chinese burrows: (i) a simple burrow, usually less than 10 m in length with two entrances (A and B), seen in cross section (*above*) and in a vertical view (*below*); (ii) cross section of another simple burrow with one entrance (A), an air hole (B), and a widened section of the tunnel (C); (iii) a complex tunnel system that can be up to 25 m long. This burrow has three entrances (A, B, C) and a terminal chamber (F). Small side tunnels (D, E, G) were dug by juvenile alligators. The burrow is seen in cross section (*below*) and vertically (*above*). Modified from B. Chen et al. 2003.

suggesting that alligators continue to excavate throughout their lives. The multilevel structure of some of the dens may be an adaptation to variability in the depth of the water table (Chu 1957). In some of the hillside ponds where water levels vary drastically throughout the year, we observed entrances to burrows that were 50 m or more from the nearest water. In certain instances it was apparent that the alligators excavated their burrows in areas where the water table was near the surface so the burrows would retain water even if the water level in the nearby ponds was greatly reduced. Some burrows have vertical shafts that reach the surface (Chu 1957; C. Huang 1982) and may serve as ventilation holes (C. Huang 1982), allowing alligators to keep their nostrils at the surface of the water in the event of flooding. Burrows made in marshy areas in hilly terrain are less subject to flooding and are not reported to have ventilation holes (C. Huang 1982). Similar ventilation holes have also been reported from burrows of other crocodilians (Guggisberg 1972).

Adult female Chinese alligators are reported to share their dens with juveniles (C. Huang 1982). Independent digging by hatchling and juvenile alligators within these burrows may explain their more complex structure. Of 18 burrows excavated by Chu (1957), 8 were empty, 8 contained one alligator (3 females, 5 males), 1 contained 2 animals (an adult female and another unsexed animal), and 1 burrow was occupied by an adult female and 3 hatchlings. In the latter case the hatchlings occupied small side tunnels that had been excavated off the main burrow (Figure 5.7iii). Hsiao (1934) reported that among 28 alligators dug out of burrows, the only instance where there were more than two alligators in a single burrow was one that had an adult (sex not reported) and two small juveniles. In addition to providing subterranean refuges for alligators, Pope (in Schmidt 1927) found a "wild cat" in one burrow, suggesting that alligator burrows may also provide shelter for other species in the winter.

SIZE OF CHINESE ALLIGATORS

Body size has enormous repercussions for life history strategies, and smaller adult size usually translates into quicker maturation, smaller generation times, and higher potential population growth rates. Some crocodilians are true giants, with maximum lengths exceeding 6 m and body mass over a ton. While not a giant, the American alligator is among the larger species, with lengths in excess of 4.5 m (Woodward et al. 1995). The Chinese alligator, among the world's smallest crocodilians, is approximately half the length of the American species. Like all reptiles, crocodilians have indeterminate growth (they continue growing as long as they live) because the epiphyses in their long bones never ossify (de Ricqles 1975). Nevertheless, growth slows down greatly after sexual maturity and in large, old individuals can be insignificant (Tucker et al. 2006). Based on growth curves calculated from captive individuals at the ARCCAR, R. Wang et al. (2006) found that growth would be negligible in females once they were 25 years old and males when they reached 35.

Size comparison for different organisms is not as easy as it might appear. Adult size can be measured as size at maturity, some sort of average size, or as a frequency distribution of sizes. For most crocodilians, particularly the largest species, a great deal of attention has been focused on maximum length. But a measure of the mean or modal length of adults is more meaningful in a biological sense. When discussing size, the sex of individuals is an important consideration as all crocodilians are sexually size dimorphic, with males usually growing approximately 20% longer than females. The size difference between the sexes is attributed principally to a slowing of female growth when they attain sexual maturity (Jacobsen and Kushlan 1989; Tucker et al. 2006) and the investment of more energy in reproduction. Larger male size is a fundamental component of crocodilian life history, with an advantage for large males when competing for females in a polygynous breeding system.

Early reports suggested Chinese alligators were quite diminutive for crocodilians. Based on his examination of a relatively small sample of animals, Fauvel (1879) considered 1.75 m to be the greatest length attained by Chinese alligators, although he surmised that larger individuals might exist. Hsiao (1934) reported that the largest captured alligator from a sample of 28 wild individuals was 96 in (2.43 m) long, but this appears to be a typographical error as later he notes the same animal was 69 inches (1.75 m) long and that the maximum size of the species was 71 in (1.80 m). In the 1950s, the largest individual in a sample of 39 wild-captured alligators was a female 1.67 m long (Chu 1957; Figure 5.8). The Anhui breeding center was established in the early 1980s by capturing 212 individuals, by far the largest sample of wild alligators, with the largest individual measuring 1.96 m TL (C. Wang, pers. comm.). More recent reports confirm that the maximum size of Chinese alligators in the wild is approximately 200 cm TL. The largest of 7 adult males collected in the wild in Zhejiang Province was 203 cm TL (M. Huang et al. 1987). In May 1995 the capture

BOX 5.3

Alligator Holes in the Everglades

Alligators are emblematic of the Everglades, arguably the most famous wetland in the world. But many parts of this enormous wetland system in southern Florida provide only marginal habitat for alligators. One reason why alligators are so renowned in the Everglades is the tendency to dig and maintain holes that serve as vital dry-season refuges for a variety of wetlands fauna. The gator holes of the Everglades are the best-known evidence of crocodilians as "ecosystem engineers": they increase spatial hetero-geneity in the wetland, promote the growth of certain important veg-etative communities, and overall provide a boost to the biodiversity in the surrounding area (Palmer and Mazzotti 2004). The Everglades holes are true holes, but some also have burrows that extend 3 to 5 m in from the edge of the hole, usually under the root mass of willow trees (Hines et al. 1968). Alligators create holes in many seasonally fluctuat-ing wetlands, but they have been best studied in the Everglades.

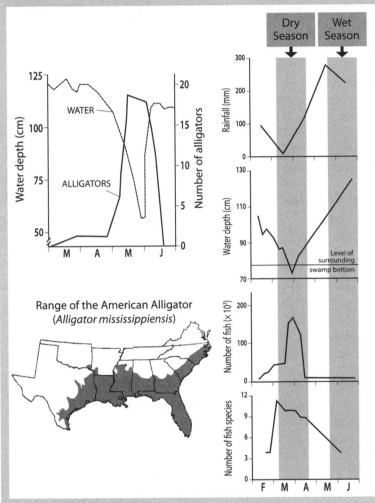

The ecology of alligator holes in the Florida Everglades.
Graphs modified from Kushlan 1974.

Initial reports of the ecological importance of these gator holes were largely anecdotal in nature (Craighead 1968; Carr 1963). Kushlan (1974) was the first to quantify the ecological value of these holes, not in the Everglades but in the adja-cent Big Cypress Swamp. In the more northern latitudes of Louisiana, cold winter temperatures, and perhaps the presence of deeper sediments, alliga-tors commonly dig bur-rows (long tunnels) into the sides of the gator holes into which they retreat during the winter. Alligator holes in Louisiana and the Everglades play an important ecological role as aquatic refuges for fish and other wetlands fauna during periods of low water (Kushlan 1974). Sediments build up around the holes that the alligators excavate. Aided by incidental fertilization of the soils by many birds and mam-mals during low water periods, the result is distinctive and more vigorous growth of vegetation that contrasts with that of the surrounding marsh. Some studies in the Everglades have taken a detailed look at the structure, function, and ecology of these gator holes (Campbell and Mazzotti 2004; Palmer and Mazzotti 2004).

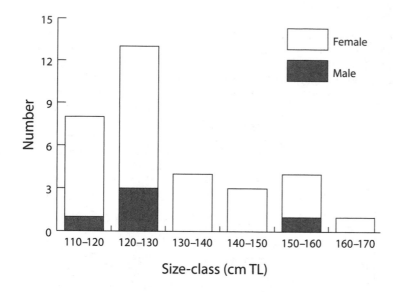

Figure 5.8. Size-class distribution of 33 wild-captured Chinese alligators over 110 cm TL. Six individuals less than 110 cm from the original sample have been omitted. From Chu 1957.

of a 1.98-meter male (38 kg) by farmers near the city of Ma'anshan (Anhui Province) was widely reported in local newspapers.

In captivity, Chinese alligators appear to grow larger than their counterparts in the wild. Among the largest captive Chinese alligators were a 1.99-m male housed in the Bronx Zoo for several years; a captive specimen in Russia measured 2.03 m and weighed 42 kg (J. Behler, pers. comm.). A male at the National Zoo in Washington, D.C., measured 2.01 m TL when it died in 1976. In China, M. Huang et al. (1987) reported a male of 2.03 m TL in the Hangzhou Zoo., Captive individuals at the ARCCAR breeding center in Anhui Province are by far the largest *A. sinensis* reported. Growth models based on measurements of ARCCAR animals suggest that mean maximum asymptote for male size is 219 cm TL (R. Wang et al. 2006). Adult males over 220 cm TL are not unusual, and the largest individual was 246 cm TL and weighed 84 kg (C. Wang, pers. comm.).

While the largest individuals in the wild reach a maximum length of approximately 200 cm, the mean size of adult male Chinese alligators is harder to quantify in the absence of an easy way to recognize sexually mature individuals. Size of animals almost certainly will vary from population to population depending on environmental, and possibly genetic factors. Of the 33 wild-captured alligators recorded by Chu (1957), only 5 were males and all were fairly small (maximum 156.8 cm TL). However, of a sample of 7 adult males captured by M. Huang et al. (1987) in Zhejiang Province, the mean total length was 158.2 cm (±26.0 cm).

In the past the largest adult females did not exceed 175 cm TL (Huang, in Groombridge 1982), and this record was probably based on the report of Hsiao (1934) of a wild captured female that was 5 ft 9 in (175 cm) long. However, under the right conditions captive animals can grow larger. The largest female at the ARCCAR was reported to measure 207 cm TL and weigh 50 kg (C. Wang, pers. comm.). Growth models based on measurements of females from this same population suggest the mean female maximum size asymptote is 173 cm TL (R. Wang et al. 2006).

The minimum size of reproduction in females was reported by Huang (in Groom-bridge 1982) to be 92 cm TL. However, this appears to be an erroneous figure from Chu (1957) based either on a female missing part of its tail or mismeasured (using the accompanying head length value the female measured 116.6 cm). The minimum total length for female reproduction reported by B. Chen (1985) was 110 cm. Using this value as a minimum reproductive threshold, the mean size of a sample of 24 adult females (> 110 cm TL) captured by Chu (1957) was 129.8 cm TL (range 114.5–159.5) (Figure 5.8). M. Huang et al. (1987) report an average adult female size of 131.1 cm TL (range 116–150 cm, n = 7). B. Chen (1991) gives an average female size of 135.0 cm (117.5–144.0, n = 19), and B. Chen (1985) estimates that females become mature in 6 to 7 years. Two adult females used to establish a breeding program in Louisiana measured 168 and 159 cm (Joanen et al. 1980). The age of maturity in the wild is not known but in captivity both males and females can reproduce at 7 or 8 years (K. Shi et al. 2006), similar to the age of maturity reported for wild American alligators (6–8 years) in coastal Louisiana (Chabreck and Joanen 1979) but considerably less than at more northern latitudes (15–18 years in North Carolina; Klause 1983).

GROWTH RATES

How fast alligators grow is one of the most fundamental characteristics of their life history. Within a population, differences in growth rates and adult size are influenced by a number of factors, including the abundance and nature of available prey, temperature, and seasonal fluctuations in water levels (Tucker et al. 2006). Social factors such as population density likely also play an important role (Lang 1987). Over the last few decades it has also become apparent that conditions experienced by embryos within the egg during incubation can have a significant influence on posthatching growth and survivorship of reptiles with temperature-dependent sex determination, which includes all crocodilians (Webb and Cooper-Preston 1989; Lang and Andrews 1994; Lance and Bogart 1994).

Information on the growth of Chinese alligators comes principally from studies of captive animals. Because they are fed regularly, alligators in captivity are likely to grow more rapidly than in the wild. But growth of captive animals can vary tremendously depending on various conditions. Most studies of growth of wild crocodilians have been done by marking and recapturing individuals and calculating the growth rate, either in terms of body length or total length, during the intercapture interval (Tucker et al. 2006). An alternative method is modeling growth based on the size of known-age individuals (Brisbin 1990). A few studies have attempted to estimate long-term growth by aging animals using histological sections of bones (Hutton 1986; Woodward and Moore 1992), but the results have been equivocal largely due to extensive remodeling of bone and the subsequent loss of early growth rings as animals age.

Not unexpectedly, studies suggest that Chinese alligators grow slower than American alligators. Herbert et al. (2002) kept hatchling American and Chinese alligators under identical conditions in a laboratory and reported that American alligators grew much faster (mean ~120 cm TL after one year) than the Chinese alligators (mean ~ 60 cm after one year). The growth rates of hatchling Chinese alligators reported by Herbert et al. (2002), while slow, were much higher than those found by Z. Zhang et al. (1985) at the ARCCAR, who reported growth of only 5 to 6 cm in the first 260 posthatching days in captivity. That study, however, was conducted during the early years of captive breeding when husbandry protocols were not yet well established. Other reports from the 1980s suggest similar slow growth, with captive animals in China reaching an average of 37 to 38 cm TL after one year, and 47 to 49 cm TL after two years, with the largest individual measuring 64.5 cm TL after its second winter (Z. Zhang and Wang 1987).

More recent studies of captive alligators have included information on the growth of large juveniles and young adults. The most extensive study of the growth of captive Chinese alligators was done at the ARCCAR by R. Wang et al. (2006) based on the measurements of several hundred known-age individuals of both sexes. Male and females grew at equivalent rates until the fourth year of life, when female growth slowed proportionally more (Figure 5.9). Overall, alligators showed annual growth increments of 17 to 25 cm during their first 4 years, slowing to 11 to 14 cm per year up to 10 years of age in males, and 8 to 11 cm per year in females over the same period of time. Animals in the 10– to 15–year range grew much slower; 4.7 cm per year in males and 1.1 cm per year for females.

Under natural conditions or the seminatural conditions of Chinese alligator farms, growth was evident only during the time of year when alligators were active. Hibernating animals did not change size significantly. Nevertheless, Herbert et al. (2002) found an interesting difference in the seasonal growth patterns of the two species. When kept at a constant temperature (31°C) throughout the year, American alligators grew at a fairly constant rate but the Chinese alligator growth rate dropped off in the winter, apparently as a result of shortened day length. Whether this represents real differences in the two species' seasonal growth patterns warrants future investigation.

The one attempt to evaluate the growth of wild Chinese alligators was carried out in the 1930s by Hsiao (1934) based on a growth curve for alligators constructed from the number of growth rings (assumed to be annuli; see "Maximum Age" below) on the scutes of alligators dug out of winter burrows. The author indicated that the sample contained 23 animals, but he presented data from only 8 females (Figure 5.10), so this may be a biased representation of his sample. While the results of the study were in general agreement with reported growth of captive animals, the validity of using external osteoderm growth rings as annuli has never been tested on Chinese alligators nor any other crocodilian.

Growth is one of many aspects of alligator biology that is influenced by environmental temperature. Alligators stop feeding and become dormant when environmental temperature drops to about 16°C (Lance 2003). Growth rates have been thought to be proportional to the length of the annual activity period, and in American alligators there is a tendency for average growth rates to decline as one moves farther north (Deitz 1979; Klause 1983; Fuller 1981). This is an important correlate for several life history traits as sexual maturity in alligators appears to be related more to size than to age (Wilkinson and Rhodes 1997). Female alligators throughout their range become sexually mature when they are about 1.8 m TL, which can come in as little as 8 to 13 years in Louisiana (Rootes et al. 1991) or as long as 15 to 19 years in North Carolina (Klause 1984).

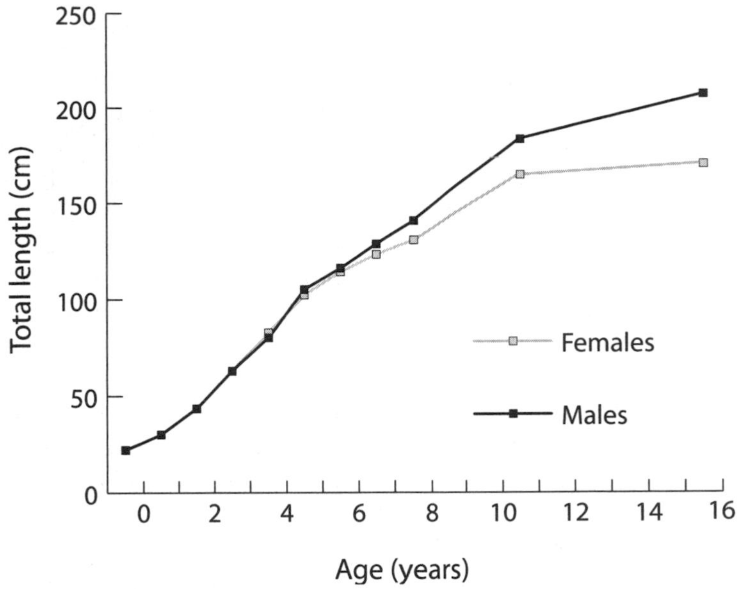

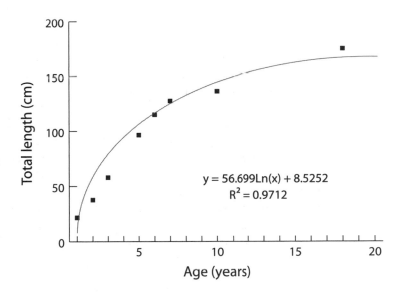

Figure 5.9. Growth curves for male and female Chinese alligators based on measurements of captive animals at the ARCCAR. Calculated from R. Wang et al. 2006.

Figure 5.10. Growth curve for female Chinese alligators calculated by Hsiao (1934) based on presumed annuli in osteoderms.

$$y = 56.699\mathrm{Ln}(x) + 8.5252$$
$$R^2 = 0.9712$$

The small size of hatchling crocodilians is another important life history trait. All crocodilians hatch at a very small size relative to adults, but in terms of absolute size, hatchling Chinese alligators are the smallest of all the crocodilians. Furthermore, after hatching, neonate alligators have a growing period of only about 2 months before ambient temperatures become so low that they stop feeding. Hatchlings measure 20 to 22 cm long and weigh approximately 25 to 30 g when they emerge from the egg, and they only grow to approximately 25 to 26 cm TL and 50 to 55 g before their first winter (B. Chen 1985; Z. Zhang and Wang 1987). At this size they are extremely vulnerable to predators and unable to excavate burrows deep enough to survive the winter. They are, therefore, very dependent on their mothers through their first winter.

MAXIMUM AGE

Fauvel (1879) noted that when the Chinese wanted to indicate that a person was especially old they would say he or she was "older than the T'ó," and that if Westerners attributed 9 lives to cats, the Chinese should give at least 12 to the alligator. While crocodilians are assumed to be among the longest-living reptiles, we know very little about their actual lifespan. And, as with size, the discussion of age usually centers on maximum values. As crocodilians are characterized by high natural rates of mortality among young animals, mean lifespan has little value (other than to calculate survival rates). What we really want to know is once alligators reach adulthood, where they have few natural predators and mortality rates are usually low, how long can they live? Two sources of information have been used to get at this question. The first is records of animals in captivity. The second is histological analysis of the bones of wild-captured animals and the identification of LAG (lines of arrested growth), which in crocodilians inhabiting seasonal environments are equivalent to growth annuli. Bone cross sections can be taken from long bones of a dead animal or from dermal bones (e.g., nuchal osteoderms) of a live one. Skeletochronology has been attempted in a preliminary fashion on American alligators (Woodward and Moore 1992), and more extensively in Nile and Australian freshwater crocodiles (Hutton 1986; Tucker 1997). The oldest individuals in a sample of 23 large Nile crocodiles were two females estimated to be 63 and 64 years old (Hutton 1986). This technique has its limitations as a result of extensive remodeling of bone structure, particularly among females who use their bones as a source of calcium for egg shell production (Wink and Elsey 1986), and ages of animals more than 10 years old may not be accurate.

Maximum age reports for alligators are based entirely on captive animals. The oldest confirmable age of an American alligator is one that died in Paris at the age of 85 (Pellegrin 1937). Pope (1940) reported that Chinese alligators in Germany had been in captivity for 50 years. Behler (1993) noted that the oldest Chinese alligators in captivity were believed to be 60 to 70 years old, and that healthy animals have continued to reproduce into the fifth decade of life. The oldest reproductive female on record was sent to the National Zoological Park in Washington, D.C., in September 1937, where she remained until 1976 when she was used in the newly established breeding program at the Rockefeller Wildlife Refuge in Louisiana. She nested in 1980 when at least 43 years old, producing 8 live hatchlings from 22 eggs (Davenport 1982). At the ARCCAR breeding center in Anhui, 64 (of 212) wild alligators captured beginning in 1979 were still alive in 2006 (J. Zhu, pers. comm.). As many of these were reported to be adults, and assuming they were at least 20 years old when they came to the ARCCAR, some of these animals would be a minimum of 47 years old.

REPRODUCTIVE BIOLOGY

Studies of the reproductive biology of Chinese alligators began in earnest with the captive breeding centers established in 1979. Prior to that, there was some information available from captive animals in U.S. zoos, as well as some largely anecdotal information in general treatments of Chinese alligators such as B. Chen (1985). But prior to our work in the late 1990s there had been no systematic attempts to gather data on the nesting ecology of wild Chinese alligators. This stands in sharp contrast to the American alligator, for which fairly detailed studies of reproductive ecology had been carried out in a variety of habitat types. Here we summarize current knowledge on Chinese alligator reproduction and try to place it into ecological context by referring to what is known about American alligators. We have not made an exhaustive attempt to review the reproductive biology of the American alligator, but present some of the more interesting recent findings that help us understand the question of how Chinese alligators produce offspring.

Reproductive Strategy of Alligators

As is typical of long-lived animals, the onset of sexual maturity of Chinese alligators is delayed and reproduction continues for decades. However, unlike birds and mammals that share this reproductive cadence, crocodilians produce large numbers of offspring in each reproductive bout. Crocodilians are also unique among reptiles in that they exhibit a level of parental care more typical of birds and mammals (R. Shine 1988), and the combination of these characteristics results in a distinctive suite of life history traits. Crocodilians are sexually size dimorphic, with males growing larger than females. One reason for this is that sexually mature females divert significant amounts of their annual energy budget to reproduction instead of growth. The investment of these finite energy resources in reproduction can be measured by four principal parameters: the size of eggs, the number of eggs in a clutch, the frequency of clutches, and the amount of energy devoted to posthatching care. In an analysis of the reproductive characteristics of crocodilians, Thorbjarnarson (1996) found differences within the Crocodylia relating to the strategic investment in the first three of these (measurements of energetic investment in parental care is much more difficult to quantify). The two species of alligators stood out from other crocodilians by sharing a similar pattern of laying large numbers of relatively small eggs. The tropical alligatorids (caiman) all tended to produce larger eggs relative to female body mass. There were also differences within the caiman, with species found in seasonally variable environments (e.g., *Caiman crocodilus*) producing large clutches of average-size eggs, whereas *Paleosuchus*, which tends to inhabit much more stable aquatic environments, have average-size clutches of relatively large eggs. Thorbjarnarson (1996) concluded that large clutch sizes and higher fecundity are evident in alligatorids inhabiting environments characterized by seasonal fluctuations in temperature (*Alligator*) or water availability (*Caiman*).

Because they are long-lived, iteroparous species, crocodilians may adjust short-term reproductive output in order to maximize lifetime reproductive success, and annual nesting is the exception rather than the rule for crocodilians (Thorbjarnarson 1996). Most individual adult females appear to nest once every two years, or two years out of three. Some species of crocodile, such as the Australian freshwater crocodile, have high rates of nesting (ca. 90% of adult females nesting in any one year; Webb, Bucworth, et al. 1983), but reported values for American alligators are much lower, usually less than 50% (Table 5.1). The frequency of nesting appears to be energy-limited; well-fed individuals in captivity will nest every year. Information on the nesting frequency of wild Chinese alligators is very scarce and is based largely on anecdotal reports from sites presumed to have just one adult female, such as Hongxin, Changle, and Zhongqiao. At these sites nests have been found on nearly

Table 5.1   Population values of reproductive frequency among adult female crocodilians

Species	Reproductive frequency (% per yr)	Location	Source
Alligatoridae			
Alligator mississippiensis	68.1	Louisiana	Chabreck 1967
A. mississippiensis	63	Louisiana	Joanen and McNease 1980
A. mississippiensis	40	Florida	Guillette et al. 1997
A. mississippiensis	29	Florida	Jacobsen and Kushlan 1986
A. mississippiensis	28	Louisiana	Taylor 1984
A. mississippiensis	25.3	Louisiana	Taylor et al. 1991
A. mississippiensis	25	So. Carolina	Wilkinson 1984
A. mississippiensis	10.0–14.8	Texas	Reagan 2000
A. mississippiensis	<10	No. Carolina	Lance 1989
Caiman crocodilus	54.3	Venezuela	Thorbjarnarson 1994
Paleosuchus trigonatus	≤33	Brazil	Magnusson and Lima 1991
Crocodylidae			
Crocodylus acutus	72	Florida	Mazzotti 1983
C. acutus	63.8	Haiti	Thorbjarnarson 1988
C. johnsoni	90	Australia	Webb et al. 1983b
C. intermedius	~80	Venezuela	Thorbjarnarson and Hernandez 1993
C. niloticus	63	Zimbabwe	Hutton 1984
C. niloticus	87.6	Kenya	Graham 1968
C. niloticus	20.6	So. Africa	Leslie 1997

an annual basis, suggesting that these adult females are nesting almost every year. In captivity females have been reported to nest annually or biennially (C. Huang 1983). High rates of nesting in the wild may reflect a lack of density-dependent factors limiting reproduction as a result of the extremely low population densities or may indicate that many surviving females are old, large individuals that are more likely to nest annually. In terms of lifetime reproductive effort, alligators may compensate for lower annual investment in reproduction by having a longer reproductive lifespan, which would be the case if alligators live longer than their more tropical relatives; something that remains to be determined. American alligators nest relatively infrequently and produce a large number of relatively small eggs. This reproductive strategy appears to set it, and perhaps the Chinese alligator, apart from other crocodilians.

Courtship and Mating

Not surprisingly, the behavioral repertoire of breeding Chinese alligators is in many ways comparable to that of the American alligator (C. Huang 1982; Watanabe 1981). Both species use a similar set of visual, auditory, and tactile cues as advertisement displays and in courtship. However, there are also clear differences between the species in the use of these social signals. Crocodilians probably also make use of olfactory cues in the defense of territories and for courtship, and they have paired glandular organs on the lower surface of their jaws (mandibular glands) as well as in the cloaca (paracloacal glands). The latter produce a musklike substance that is thought to contain pheromones that play a role in reproductive behaviors (Weldon and Ferguson 1993) although these are poorly understood. In this section we describe what is known about the courtship and mating of Chinese alligators, most of which is based on observations of captive individuals and by reference to the better-studied American alligator.

Breeding Systems

The mating systems of crocodilians exhibit considerable inter- and intraspecific variability based on factors such as habitat type and population density. Among more tropical species, seasonal breeding groups form in response to dropping water levels at the beginning of the dry season. This is true for hole-nesting species such as the Nile crocodile (Cott 1961; Kofron 1990, 1993) and the Orinoco crocodile (Thorbjarnarson and Hernandez 1993), as well as some mound-nesting species such as the black caiman in the Amazonian flooded forests (Thorbjarnarson, pers. obs.). Other species, including the estuarine crocodile (*C. porosus*) in parts of northern Australia (Lang 1987) and the smooth fronted caiman (*P. trigonatus*) that lives in small, relatively aseasonal streams in the terra firme Amazonian forests (Magnusson and Lima 1991), may occupy year-round territories and not form significant breeding groups on a seasonal basis. The temperate-dwelling American alligator will form seasonal breeding aggregations in response to warming environmental temperatures in the spring (Joanen and McNease 1980).

Crocodilians in general are believed to be polygynous, with dominant males mating with multiple females. In all species of crocodilians males grow larger than females; dominance among males is hierarchical and based on largely on size. Little is known about the dynamics of reproductive territoriality, but males are known to enter into agonistic encounters with one another during the breeding season (Garrick and Lang 1977; Vliet 1989), and a dominant social status confers tremendous reproductive benefits on large males (Lang 1987). Male American alligators have larger home ranges than females (Joanen and McNease 1989; Hagan 1982). In the coastal marshes of Louisiana both sexes are reported to group in open water habitats for courtship and mating during the spring reproductive season (Joanen and McNease 1989). Radiotelemetry studies of alligator movement patterns carried out in a variety of habitat types have found that females, and in some habitats males, tend to move more during the spring mating season than at other times of the year (Chabreck 1966; Joanen and McNease 1972, 1989; Goodwin and Marion 1979; Hagan 1982), suggesting that breeding adults are seeking one another during this period. Until the advent of molecular tools based on high resolution genetic markers, the study of the mating dynamics of crocodilians was fraught with difficulties. Because courtship and mating between animals does not necessarily mean that the male has actually fertilized eggs, a thorough understanding of breeding dynamics is almost impossible based solely on visual observation. The use of genetic tools that can assign parentage is still in its infancy, and we use the results of a study on the American alligator (Davis et al. 2001; Davis 2002) to provide insights on the reproductive ecology of Chinese alligators.

The work of Davis et al. (2001) on alligator reproductive dynamics was carried out in two very different habitats—the coastal marshes of Louisiana and a freshwater pond system in South Carolina. In Louisiana adult females and offspring from

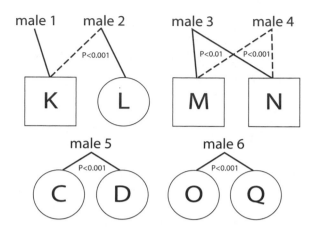

Figure 5.11. Diagram of parental relationships in American alligator nests from Rockefeller Wildlife Refuge in Louisiana. Squares denote clutches fathered by multiple males. Circles denote singly sired clutches. Bold lines indicate the primary male; dashed lines indicate the secondary male. For each pair of parental comparisons with P-values, the parental genotypes are identical across all five loci. P-values indicate the confidence that that the two identical genotypes were actually the same male. For the relationship to the primary male 3 for nests M and N, the P-value is < 0.01 due to common alleles in the multilocus genotype. Based on Davis et al. 2001.

22 nests were sampled during two separate breeding seasons. Not surprisingly, the genotypes of the hatchlings were consistent with the females at the nest being the mother. The study did not sample from all the 300 or so nests in the region, nevertheless there was clear evidence that individual males are mating with, and fertilizing the eggs of, multiple females. The genotypes of the hatchlings also confirmed for the first time multiple paternity in a wild crocodilian, with 6 of the 22 nests having two fathers, and one nest having three fathers (Figure 5.11).

This study was complemented by an examination of the breeding dynamics of alligators in a 1,100-ha freshwater pond in South Carolina over a period of six years (Davis et al. 2001). Because it was conducted in one pond over multiple years, Davis et al. (2001) were able to look at annual variability in the breeding dynamics of alligators. They found that females did not nest every year; when they did nest they tended to do so in the same general area, but not necessarily at the same site. Overall, only 22 of 54 adults (of both sexes) in the study were producing nests. Some of the females that nested multiple times during the study did so with different males, confirming the finding of multiple paternity in the Louisiana study, with 19% of the nests having more than one father. Males tended to sire nests that were in specific sectors of the pond, lending credence to the supposition that they established reproductive territories in those areas. Overall, however, there were fairly low levels of female mate fidelity over the years, suggesting that male reproductive territories, dominance status, or both may shift from year to year.

Based on what we know about Chinese alligator breeding systems, it is likely that they follow some of the same general patterns observed in American alligators. During the spring courtship and mating season males are reported to move around searching for females (B. Chen and Li 1979; Watanabe 1981). B. Chen (1985) reported that in spring males move up to 1 km looking for mates, but it is likely that in some habitats they may have moved considerably more. Outside of the mating season, Chinese alligators are more sedentary, moving only 100 to 200 m from the site of their burrow (B. Chen 1985; B. Chen et al. 2003), but this very much depends on the nature of the habitat. In response to flooding or drought or if their habitat is disturbed or destroyed, alligators are capable of moving much longer distances. Two individuals that escaped from breeding centers were recaptured 20 and 32 km away (B. Chen 1985).

In captivity, males engage in combat and in a variety of stereotyped postures and movements that play a role in establishing social hierarchies. Males have been observed repeatedly swimming or walking parallel to one another in inflated postures for several hours (K. Vliet, J. Brueggen, and J. Thorbjarnarson, pers. obs.). This may allow individuals to gauge the other's body size and determine social position without resorting to potentially injurious physical combat. The female-biased sex ratios of adult animals found in winter burrows may be evidence of a system based on male hierarchical dominance and polygyny. A group of 23 animals found in burrows by Hsiao (1934) had a sex ratio heavily biased toward females (3 males, 20 females; 1:6.67). "As our collections were made at random it seems likely that in nature females are more numerous than males" (Hsiao 1934). A similarly female-biased sex ratio (1:5.5) was reported by Chu (1957) based on a sample of 39 alligators taken from burrows. If these were individual breeding colonies of alligators, the paucity of adult males may indicate that the socially dominant males exclude smaller conspecifics. If this is the case, then the smaller adult and subadult males may have been forced to occupy peripheral areas and overwinter in burrows removed from the main breeding groups.

Vocalizations

Crocodilians, which are among the most vocal of reptiles, have a sophisticated repertoire of vocalizations that play an important role in a variety of behaviors, including courtship and presumably establishing and maintaining reproductive territories (Garrick and Lang 1977; Lang 1987). The Chinese alligator is a particularly vocal crocodilian; it makes a variety of sounds ranging from the grunt and "distress calls" of hatchlings and juveniles to the roars or bellows produced by adults as advertisement displays. As with other crocodilians, hatchling and juvenile alligators emit high-pitched gruntlike vocalizations that serve to communicate with other juveniles as well as adults (H. Campbell 1973; Herzog and Burghardt 1977). Full-term embryos will give these "hatching calls" from within the egg as a signal for the female to open the nest. Vocal signals are a particularly important link between adult females and their offspring and to maintain group cohesion among neonates (contact calls). When confronted by a threat, juveniles will give a louder "distress call," which presumably serves to alert other juveniles as well as to attract the attention of an attending adult. The acoustic structure of all these juvenile calls is similar but differs in terms of volume, context, and cadence (Britton 2001).

Observations of captive Chinese alligators in the Changxing breeding center

Figure 5.12. *Clockwise from top left*: Jawclap sequence of an adult male Chinese alligator.

(K. Vliet, J. Brueggen, and J. Thorbjarnarson, pers. obs.) suggest that adult Chinese alligators are also highly vocal and visually oriented in their behaviors during the mating season. Both male and female Chinese alligators produce a graded series of advertisement displays that incorporate both vocal and nonvocal auditory signals. The preliminary descriptions presented here are based on less than 10 hours of observation at Changxing during the May 2007 breeding season.

Chumpf Bellow (Burp Bellow)

This display, produced by both males and females, is a combination of signal elements involving body posture and vocal (chumpf bellows) and nonvocal (headslaps) auditory signals, usually with the alligator's head in shallow water near the shore and body extending perpendicular to the shoreline. First, the alligator swims slowly toward the shore in an elevated posture, with its head, back, and tail partially out of the water. Prior to the headslap the alligator assumes a HOTA posture (similar to behavior seen in American alligators; Vliet 1989) or lifts its body partially out of the water on its limbs ("inflated posture"; Garrick et al. 1978), increasing the distance between the jaws and the water, which presumably allows a more forceful display. The auditory signals begin with one or two forceful headslaps, a sudden slap of the undersurface of the alligators head against the water (Figure 5.12). Headslaps are accompanied by jawclaps, a quick opening and closing of the jaws that produces a loud percussive sound (Herzog 1974; Vliet 1989). The headslap-jawclap display is followed by a series of "chumpf" bellows, vocalizations that lack the strength of the full bellow and are produced from a HATO body posture. After the vocalizations the alligator lowers its

Figure 5.13. Body posture of a male Chinese alligator during a bellow: (a) raised head–oblique tail arched posture and inhalation; (b) dropping down and producing the audible bellow and subaudible vibration (visible as agitated water over the alligator's back); (c) rising up again onto extended front legs to repeat the bellowing cycle.

stance in the water, with only the head and part of the back exposed, and produces a series of low-intensity jawclaps, giving the impression that the animal is biting the water. As these water bites decrease in intensity, the alligator drops down even farther in the water and ends the display by exhaling air through its mouth, producing bubbles. Low-frequency subaudible vibrations are produced by male American alligators in association with the headslap display (Garrick and Lang 1977; Vliet 1989) but we did not observe it in this context with Chinese alligators.

Figure 5.14a–c (*facing page and above*). Stereotypical body posture of Chinese alligators during a bellowing chorus.

In American alligators the social function of the headslap display has been suggested as a declaration of presence, that is, as a means of alerting other alligators to the presence and the location of the animal conducting the display (Garrick and Lang 1977; Vliet 1989).

Bellow

Both adult male and female Chinese alligators produce a loud, explosive series of calls that are believed to play a role in reproduction by attracting individuals of the opposite sex, spacing out animals, or both (Vliet 1989). Many crocodilians are known for these loud vocalizations, usually referred to as bellows, that are characteristic, but variable, combinations of vocal signals and typical body postures (Garrick and Lang 1977). The bellowing behavior of American alligators has been well described (Garrick and Lang 1977; Vliet 1989) and serves as a point of reference for our observations on Chinese alligators. A bellow is a "repeated alternation of stereotyped postural attitudes coordinated with inhalation and exhalation" (Garrick et al. 1978). The vocalizations, which are made during the exhalation, are referred to as the *bellow*, and the elements that are repeated from bellow to bellow are a *bellowing cycle*. The repeated bellows of one individual are called a *bout*. More than one alligator bellowing simultaneously is referred to as a *chorus*.

Alligators bellow when in shallow water, and they begin from an elevated posture. Adult males will raise their snout and tail to enter a HOTA body posture similar to that used in the chumpf bellow–headslap. The bellowing cycle begins with the alligator extending its forelimbs to raise its body and at the same time shifting its body

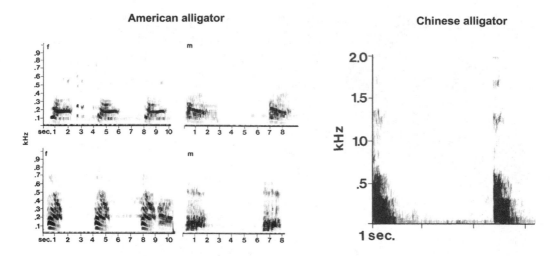

Figure 5.15. Sound spectrograms of four American alligators (2 male, 2 female) and an adult male Chinese alligator. Modified from Garrick et al. 1978.

slightly forward and inhaling (Figure 5.13a). Inhalation is visible by the extension of the alligator's gular region as it appears to be gulping air. When the head is raised, the tail is also lifted, in part to counterbalance the raised forequarters, and in small males and females this can result in the distal half of the tail being lifted completely out of the water (Figure 5.14a–c). From this position, the alligator drops its forefront down into the water, moving slightly backward, and produces an audible bellow while exhaling. At this time the tip of the snout is extended even more, resulting in the head being elevated at a 40- to 50-degree angle (Figure 5.13b). A brief subaudible vibration, visible as a "water dance" (Vliet 1989) of agitated water over the back of the male alligator (Figure 5.13b), is given at the same time as the audible bellow. If the alligator is going to bellow again the bellowing cycle then starts anew.

Garrick (1975) studied bellowing of adult male and female Chinese alligators in U.S. zoos. He noted that both males and females produce bellows, and they did so from a typical HOTA posture similar to that observed in American alligators. Bellowing usually occurred in bouts of 2 to 8 consecutive bellows, with most of the sound energy in the 25 to 500 Hz range (Figure 5.15). He also noted that there were sex- and individual-specific qualities to bellow structure and pattern, that bellowing was more common during the spring courtship season, and that courtship was frequently preceded by bellowing.

While the behavioral and postural sequence of bellowing is broadly similar in the two species, the structure and pattern of American and Chinese alligator bellows are different (Garrick et al. 1978). Chinese alligator bellows are best described as short, explosive roars; they are the principal vocalization for long-distance communication among alligators. Mean sound energy of Chinese alligators bellows is 90 dB at a distance of 5 m (X. Wang et al. 2007). The sound is not unlike the Chinese word *T'o* (the old term for Chinese alligators) if it is emphatically pronounced, carefully enunciated, and slightly drawn out (Brazaitis 1968). Garrick (1975) likens it to the sound made by "striking the closed end of a standard sized metal garbage pail." Male

Chinese alligators tend to have more bellows per bout (Garrick et al. 1978) but the studies conducted have not included sufficient individuals to allow an analysis of other sex-related differences.

The bellows of American alligators are produced in association with typical body postures and have been described as a "booming roar" (Garrick et al. 1978). The seasonal pattern of bellowing is similar to that of the Chinese alligator, with bellows being heard throughout the year but peaking during the courtship and mating season. American alligators typically bellow 4 to 6 times per bout and each bellow is longer in duration than that of the Chinese alligator. There are no sex-related differences in the number of bellows per bout (Garrick et al. 1978; Vliet 1989), but there are very clear differences in the vocalizations made by male and female American alligators. The bellow growl is produced by female American alligators, usually in response to nearby agonistic encounters among other alligators or when threatened by a large male (Garrick et al. 1978; Vliet 1989).

Males of many crocodilians produce very low frequency infrasonic sounds that are below the range of human hearing. The production of these sounds is evidenced by the agitation of the water on the back or around the torso of the vocalizing individual. In American alligators Vliet (1989) found that only males produced an infrasonic subaudible vibration. This is done immediately prior to the audible bellow, whereas in Chinese alligators it is produced at the same time as the audible bellow. The interbellow interval in American alligators is shorter and the cadence of bellowing bouts is also faster in females, mostly as a result of their not producing the SAV (Vliet 1989).

Bellows are contagious, with the bellow of one animal stimulating others nearby and can result in extended bellowing choruses by a number of individuals, particularly in captivity. A study of bellowing behavior of alligators at the ARCCAR breeding center by X. Wang et al. (2006) found that bellowing choruses were more frequent during the spring mating season (4.69 choruses per day) than other times (1.77 choruses per day) and that more alligators participated in the bellowing choruses during the mating season (28.6 alligators per chorus versus 16.5 per chorus at other times of the year). The length of bellowing choruses was similar during both periods (mean ca. 10 minutes). Bellows by Chinese alligators at the ARCCAR frequently result in animals moving from land into the pond, particularly during the mating season (X. Wang et al. 2006, 2007). In American alligators, opposite gender pairs are known to pair up in close proximity to each other for bellowing (Garrick and Lang 1977; Vliet 1989). Similar observations have been made in Chinese alligators (Figure 5.15). In some circumstances this behavior devolves into courtship and mating (Garrick and Lang 1975; Vliet 1989).

Alligators bellow during most of the year while they are active, with a peak during the mid-June courtship and mating period (Watanabe 1981; B. Chen and Li 1979). We studied the seasonal and daily patterns of bellowing in a wild group of alligators at Hongxin pond from 2000 to 2003 (J. Wu and Wang 2004). This pond had a small population of wild alligators, and a maximum of three alligators were heard to bellow at any one time. It was not possible to identify individuals or the sex of bellowing individuals, but these data provide a good overview of the seasonal and daily pattern of bellowing. Observations from 2000 to 2002 were taken opportunistically; in 2003 sampling was done in a more systematic fashion (Figure 5.16). Bellowing bouts

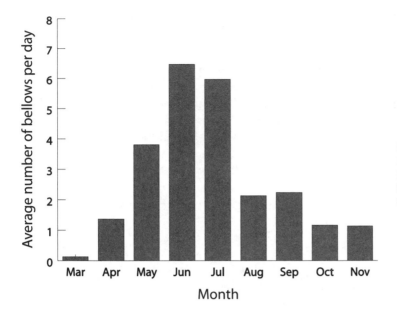

Figure 5.16. The average number of bellowing bouts per day by month at the Hongxin pond in 2003. From J. Wu and Wang 2004.

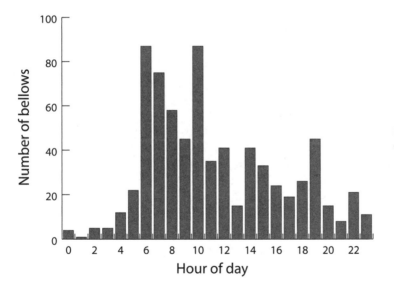

Figure 5.17. Diurnal patterns of observed bellowing bouts at Hongxin 2000–2003.

(a series of bellows by one individual) and choruses (bellowing by more than one individual) began in late March (20–26 March) and peaked in June and July when courtship and mating were under way. The daily occurrence of bellows tended to peak in the morning and, to a lesser extent, in the evening (Figure 5.17), with alligators bellowing 1 to 5 times per bout (Figure 5.18). Similar annual and diurnal patterns of bellowing were found by X. Wang et al. (2006, 2007) among captive alligators at the ARCCAR.

Weather influenced the frequency of bellowing. The study at Hongxin compared the frequency of bellows in each of four weather categories— sunny, partly cloudy,

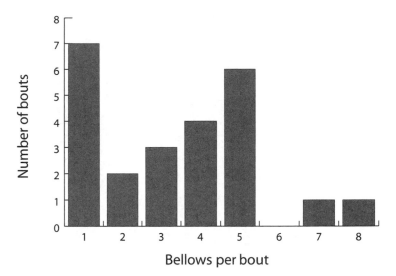

Figure 5.18. Number of bellows per bout when only one individual was bellowing.

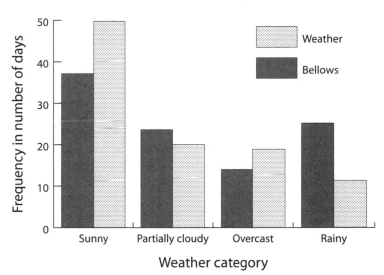

Figure 5.19. Frequency of alligator bellows by weather category compared with the frequency of each weather category. From J. Wu and Wang 2004.

overcast, and raining—to the overall frequency of each weather condition (Figure 5.19). Alligators were more likely to bellow during periods of rain ($\chi^2 < 0.001$), something that has long been recognized by Chinese farmers and may be one of the links between Chinese alligators and water deities that resulted in the development of dragon mythologies in ancient China (Chapter 4).

What purpose do bellows serve? The first is probably to attract the opposite sex for courtship and mating, and this has been mentioned by numerous authors (see Vliet 1989). Among Chinese alligators, as well as other species of crocodilians, advertisement displays are most common during the courtship and mating season. Both bellowing and headslaps will elicit approaches by members of the opposite sex in captivity (Garrick and Lang 1975). In wild populations male alligators may move long distances in search of mates and the audible signals associated with the display are likely to convey information on the location and sex of the individuals involved.

Figure 5.20 (*above and facing page*) Courtship behaviors include (a) blowing bubbles, (b) snout rubbing, and (c) parallel swimming.

Bellows may also contain information on the size and social position of the alligator that would be important for spacing out same-sex individuals and for establishing reproductive territories (Garrick et al. 1978). The sound signals from bellows, particularly the very low frequency SAV produced by males, may play an important role in long-distance communication between alligators, especially in densely vegetated habitats where visual signals have limited use (Garrick and Lang 1977; Garrick et al. 1978).

Courtship and Mating Behaviors

The courtship and mating behavior of captive Chinese alligators is similar to that described for American alligators (C. Huang 1982; Watanabe 1981). These observations confirm the important role of body-posturing, snout-rubbing, and a variety of other vocalizations described as "chuffs" or "burps." B. Chen (1985) noted that during courtship unreceptive females produce repetitive "puh-puh-puh" calls when approached by a male. Watanabe (1981) reported that immediately following a female's bellowing she and a nearby male rapidly swam toward one another and began mating. Observations of captive alligators at the Changxing breeding center suggest that both visual and tactile cues play important roles in courtship rituals. Males will swim up to females and engage in snout-rubbing, bubbling, and other tactile behaviors (Figure 5.20). One characteristic behavior we have noted among captive Chinese alligators is parallel swimming, with the male angling his neck toward the female, giving the impression of an extended stare. Males mount females in the water and arch their body and tail to oppose cloacae and achieve intromission (B. Chen 1985) (Figure 5.21).

Timing of Mating and Nesting

Most crocodilians are highly seasonal in their nesting, with ambient temperature or water levels being the two environmental cues that trigger reproduction (Lang 1987; Lance 2003). In American alligators it is the warm spring temperatures that initiate reproductive events (Lance 2003; Guillette et al. 1997). In males, reproductive status is reflected in the mass of the testes and circulating levels of testosterone in the blood. Both peak during the early spring period of courtship and mating and drop precipitously during the summer. The annual female reproductive cycle is

Table 5.2 Oviposition and hatching of wild alligator eggs (2000–2004)

Nest	Oviposition date	Hatching date	Incubation (no. of days)
ZQ 2000	02 Jul 00	29 Aug 00	58
ZQ 2002	07 Jul 02	n.d.	n.d.
HX 2002	16 Jul 02	n.d.	n.d.
HX 2003	15 Jul 03	07 Sep 03	54
ZQ 2003	10 Jul 03	05 Sep 03	57
HX 2004a	15 Jul 04	09 Sep 04	56
HX 2004b	15 Jul 04	05 Sep 04	52

Note: n.d. = no data available.

closely synchronized with that of the male.

J. Wang and Huang (1997) report that over a 12–year period at the Changxing breeding center the mating season ranged from 23 May to 27 June. Nesting takes place during a short period (usually 3 weeks) in late June and early to mid-July (B. Chen and Wang 1984; J. Xu et al. 1990; Webb and Vernon 1992). The seasonal pattern of nesting is not only constrained by temperature, but also by water levels. The warm midsummer months also coincide with seasonal peaks in rainfall and water levels (Figure 5.22). Females depend on the annual spring flush of growth in herbaceous vegetation to provide the raw material for making their nests. Unusual weather conditions may alter nesting schedules; for instance heavy rains and low summer temperatures in 1995 delayed nesting by approximately 20 days to 13 July (Z. Zhang 1995). The mean date of egg laying for 11 wild nests in Jinxiang County was 28 June, and hatching was 18 September. Dates of ovipositioning for 7 wild nests between 2000 and 2004 were all in the first half of the month of July (mean date of 12 July), with hatching for 5 of the nests in early September (mean date of 5 September) (Table 5.2). In American alligators the mean date of nesting depends on latitude, with populations farther north laying eggs at a later date (Hagan et al. 1983; Deitz 1979); the nesting dates of Chinese alligators correspond to those of the American alligator in the northern part of its distribution.

Figure 5.21. A male Chinese alligator rides atop a female before mating.

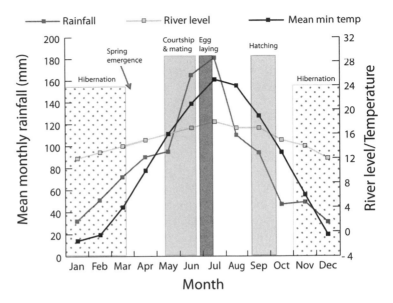

Figure 5.22. Principal annual activity periods for Chinese alligators as a function of three environmental variables.

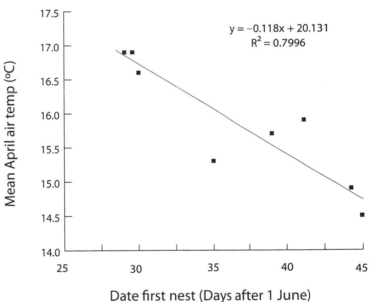

Figure 5.23. Date of oviposition for first nests at the ARCCAR as a function of mean April air temperature for the period 1992 to 2001. Alligators nest earlier following warm early spring periods. From F. Zhang et al. 2006.

Hatching dates for Chinese alligators in the NCAR (early to mid-September) correspond to that of the American alligator in the northernmost part of its distribution (North Carolina).

The annual timing of nesting is influenced by environmental factors, particularly temperature. In years with warm spring temperatures, alligators tend to nest earlier in the year. F. Zhang et al. (2006) examined the time of nest construction at the ARCCAR breeding center in relation to spring weather conditions (temperature and rainfall), and found that mean air temperature in April had the strongest affect on the date of egg laying (Figure 5.23). A similar effect of spring temperatures on the mean date of egg-laying of American alligators has been shown in Louisiana (Joanen and McNease 1980, 1989). In a subsequent analysis of nesting dates of alligators at ARCCAR, F. Zhang et al (2009) found that over the period of 1987 to 2005, mean March and April air temperature increased significantly, reflecting global warming trends. During the same period mean date of oviposition advanced 10.8 days, which the authors suggested was a response to warmer spring temperatures.

The Chinese alligator has one of the shortest incubation periods reported for any crocodilian. We were able to determine the date of oviposition and hatching of five wild nests; the mean incubation period was 56 days (range 54–58 days) (Table 5.2). The length of the egg incubation period is temperature dependent. Warmer nests have a shorter incubation period, with mean incubation periods of 64 days at 29°C, 55 days at 31°C, and 49 days at 33°C (R. Wang and Ma 2000). Other reports suggest a similar incubation period, or somewhat longer ones of 70 days (C. Huang 1982), 68 days (Z. Huang and Watanabe 1986), and 60 to 70 days (B. Chen, 1985; range 57–88 days). In a study of the embryological development of A. sinensis, Hua et al. (2004) incubated eggs at an average temperature of 29°C and found full embryonic development was completed in 56 to 62 days.

Nests and Nesting

Not all adult female alligators nest every year, but nesting, and related attending behaviors with hatchlings, has significant implications for habitat use and activity patterns. One of the most comprehensive analyses of nesting and activity patterns of female American alligators was conducted by Rootes and Chabreck (1993) using data from 15 adult females that were radio-tracked over a period of two years in the coastal marshes of Louisiana. Of these 15 females, 5 had groups of hatchlings when they were initially found, indicating that they had nested the previous year. Of the remaining 10 females, 5 nested during the first year of the study and 2 in the second. Together, these data indicate that over the three-year period an average of 30.0% of the females nested in any one year. Neonate alligators remained with their mothers for the rest of the year, overwintered with the female, and stayed together as a pod with her the following year, dispersing either shortly before or after their second period of hibernation. The adult females spent most of their time (68.6%) during the nesting season at their dens (deepened pools maintained by the alligators usually adjacent to the nest), regardless of whether they were nesting, and their mean home range size during this period was 10.2 ha; the home range averaged 25.5 ha over the entire year. Females remained close to dens during the day (87% of daytime observations were at the den), but tended to move away from them at night (46% of nighttime observations). The overall percentage of females found at their dens increased to 83.1% in the winter when they were less active. Thirteen of the 15 females changed dens during their study, but the new dens were close to the original ones and there was no discernable pattern to these changes (Rootes and Chabreck 1993).

In the floodplain marshes that were the natural habitat of the Chinese alligator it is probable that adult female A. sinensis had similar patterns of activity and nesting in the vicinity of their burrows. Among wild populations we have no information on the frequency of nesting among adult females; in captivity in Anhui it has been reported to be about 67% (Z. Zhang 1989). There is evidence that hatchlings overwinter in adult burrows. Because the Chinese alligator is a much smaller species, it is likely that home range size would be smaller than reported for American alligators. Nevertheless, the extent and seasonal pattern of movements would be expected to vary depending on the nature of the habitat.

Description of Nests

Like most crocodilians and all species in the Alligatoridae, Chinese alligators lay their eggs in small mounds of decomposing herbaceous vegetation and other organic matter scraped together by the female from the area immediately surrounding the nest (Figure 5.24a, b). C. Huang (1982) notes, "In the period between early July and late August, the female selects a secluded spot on a slope facing the sun and not far from the marshes to build a nest." In marsh-dwelling American alligators, the minimum estimated period of nest construction by females averaged 10.7 ± 4.3 days (n = 17 nests; Deitz and Hines 1980). Chinese alligators follow a similar pattern of nest building (Joanen and McNease 1981), with an average period of 10 days between initial construction of the nest mound and egg laying (F. Zhang et al. 2006). Starting about a week before building the nest mound, females begin tramping down vegetation in areas of elevated terrain near the den. To create the mound females scrape together vegetation using their front and hind limbs and also are reported to collect herbaceous vegetation in their mouths, biting it off at the roots. The female will make a small loose pile to which material is added over a period of several days and gradually shaped into a nest mound. Although similar in form to those made by American alligators, nest mounds of Chinese alligators are considerably smaller (Joanen et al. 1980). B. Chen (1985) reported the height of nests ranged from 43 to 72 cm. H. Zhu (1997) found nest dimensions at Hongxin were 45 cm high and 90 cm long.

The nest sites used by female Chinese alligators today reflect the highly altered environments in which they live; small artificial ponds in agricultural landscapes. Within these areas, female alligators seek small relatively undisturbed patches of vegetation for nesting, and the presence of sites suitable for nesting may explain why alligators have survived in certain areas and not in others. Within the ARRCAR breeding enclosures, which offer natural nesting habitat comparable to that seen in the wild, nests are located in partially shaded sites (mean 40% canopy coverage) an average of 3.1 m from the water's edge and 1.2 m above the level of the pond (F. Zhang et al. 2006).

During our 1997 and 1999 surveys in Anhui Province, we visited sites where wild nests had been reported in recent years, as well as two active nests. In these areas man-made ponds set in relict wetlands adjacent to vegetated hillsides provide a mix of habitats where reproduction and recruitment periodically occur. Active nests at two sites (Shaungken and Zhuangtou) were located at the bases of pine trees on hillsides 20 to 40 m from small artificial ponds used to store water for rice cultivation. The nest at Shaungken was situated between three pine trees, measured 90 × 80 × 40 cm high, and was composed of grass stems, leaves, pine needles, twigs, and fern fronds. The nest had a total of 17 eggs (12 banded) at a depth of 14 cm. At this site local farmers told us that in the 1970s alligators nested in a field of *Typha* situated adjacent to a small pond.

The 1999 Zhuangtou nest was on a ridge in low hills planted with pine trees above a series of small ponds. The nest mound measured 100 × 83 cm wide and 43 cm high;

Figure 5.24. Nests of wild Chinese alligators: (a) Wang Xiaoming and Ding Youzhong examine a nest in a small clearing on the island in the Hongxin pond, August 2003; (b) Wang Chaolin stands next to a nest at Zhuangtou in 1997.

it was 1 m from the base of a 4.5–m tall pine tree. The main nest material was pine needles, with small quantities of grass stems and leaves from bamboo and other herbaceous vegetation. Two nest sites from which eggs had been collected by ARCCAR were observed in this same area in 1997. One nest was at the same location as the 1999 nest. The other was on a nearby pine ridge and was qualitatively similar in terms of nest size and position. Eggs from the 1999 nest were left to incubate naturally, but none of the 19 eggs hatched (all eggs were banded), presumably as a result of low incubation temperature.

Local farmers report that alligators also use small islands for nesting. Like the vegetated hillsides, small islands set in agricultural ponds receive relatively little human attention and offer terrestrial refuges where females can build nests. At Huangu village in Guangde County, nesting was reported to have occurred during the 1960s on an island (5 m in diameter) in a small lake. B. Chen (pers. comm.) also found nests in rice fields and along small streams in hilly areas in the 1970s and 1980s.

After 1999 we were able to collect additional information about nests at two sites, Zhongqiao and Hongxin. In all cases nests were constructed from the available leaf litter mixed with herbaceous vegetation and shaped into a mound 96 to 130 cm in diameter and 30 to 43 cm high. Nests at Zhongqiao were located on a heavily vegetated dam that creates the principal pond inhabited by the alligators, under a dense canopy of small trees and bamboo. The nest mounds were constructed largely of bamboo leaves, with other leaf litter material intermixed. Nests at the Hongxin site were made on the small island in the pond and were composed principally of herbaceous vegetation from the area immediately surrounding the nest mound. In 2004 and 2005 two nests were found on the island.

The locations where alligators nest in China today are highly biased and principally reflect the use of relatively undisturbed areas that are adjacent to the last remaining, highly modified habitats where the alligators survive. Up until the twentieth century, the limited biological work on alligators was carried out almost exclusively during the winter, when the marshes were dry and alligators were easily located in their burrows. As a result there are no early accounts of the natural history of Chinese alligators during the summer, when alligators nested, and there is little to go on in terms of reconstructing the habitats where Chinese alligators nested under more natural conditions. Based on what we know about the nesting ecology of other crocodilians in seasonally flooded marsh habitats, we surmise that alligators nested in one of two general habitat types: on floating mats of vegetation or on the highest points in the uneven terrain of the floodplains. These latter areas would also provide alligators with the opportunity to dig burrows that could be used as retreat sites during the summer. A close reading of the description of the burrows in Hsiao (1934) and an examination of the figures in Chu (1957) (particularly the adult female burrow) suggests that at least some burrows were located in such elevated microsites. Pope's description of alligator burrows being found in distinct "colonies" in the marshes would also be consistent with this interpretation.

Egg Laying and Hatching

Females lay their eggs in holes they dig in the top of the nest mound, covering the nest cavity with material from the mound when they are finished. At the Shanghai

Zoo, C. Huang (1983) noted that oviposition lasted 30 to 40 minutes and took place at dawn. K. Shi et al. (2006) reported that oviposition lasted from 30 minutes to 3 hours.

In all crocodilians that have been studied, the mother plays an active role in the egg hatching process (R. Shine 1988). Females remain in the vicinity of their nests during the incubation period and may keep egg predators away, although today there are few natural predators of Chinese alligator nests. The females regularly visit their nests toward the end of the incubation period and respond to vocalizations of the young by excavating the nest. If necessary, the mother assists neonates by gently breaking open the egg in her mouth, and carries neonates in her gular pouch to nearby water (Z. Huang and Watanabe 1986). In most cases if females did not open the nest mound hatchlings would be unlikely to dig their way out of the dried and compacted nest mound without assistance.

Z. Huang and Watanabe (1986) provide an overview of nest-opening behavior.

> In response to the chucks from the unhatched eggs, which may be heard 1–2 days before the breaking of the egg shells, the female digs up the nest with its fore-limbs to release the young. If any hatchling is not powerful enough to break out of the shell, the mother will help it to do so, using [her] jaws. The first author witnessed the death of hatchlings which had got entrapped in entwined grass in the absence of the female. If the hatchlings are agitated and call in alarm, the mother will pick them up in her mouth and take them to a safer place. Young usually go into the water at a 2–day age, lured by their mother's vocalizations. . . . Females of both species [American and Chinese alligators] exhibit removal of egg debris from the nest areas during and immediately after excavation, which serves to remove a possible source of chemical and visual attraction to a potential predator.

Posthatchling maternal care in American alligators has been well documented (Hunt 1987; Kushlan 1973), and anecdotal evidence suggests that female Chinese alligators likewise remain with hatchlings during their first few months and that hatchlings overwinter in the female's den. K. Shi et al. (2006) note that juveniles will overwinter with females until they are 2 to 3 years old and capable of digging their own burrows.

Egg, Clutch, and Hatchling Characteristics

Another key set of reproductive parameters for understanding the population dynamics of alligators and how they may differ from other species are the characteristics of eggs and clutches. Relative to body size the two species of alligator produce large clutches of small eggs when compared with other crocodilians (Thorbjarnarson 1996). There is also evidence that alligatorids tend to reproduce less frequently (e.g., a smaller proportion of adult females nest in any one year) than do crocodiles.

In real terms body size is a critically important variable, and allometric trends in reproductive characteristics are an essential life history feature in crocodilians (Thorbjarnarson 1996). For any species, female fecundity is a product of two factors: the frequency of reproduction and clutch size. In most cases both these parameters tend to increase with female body size, demonstrating a clear size-specific trend in fecundity. Crocodilians have nondeterminant growth, and females continue growing, albeit at a low rate, throughout their entire lives. Consequently, a general characteristic among crocodilians, and one that is intimately linked with their strategy of long lifespan and iteroparous reproduction, is that older females, by virtue of their larger size, are more fecund than younger individuals.

Egg Size

Chinese alligators lay the smallest eggs of any crocodilian (Thorbjarnarson 1996). Mean egg mass for *A. sinensis* has been reported as 51.7 g (Joanen et al. 1980), 44.6 g (C. Huang 1982), 45 g (H. Zhu 1997; mean dimensions 3.5 × 6 cm), and 42.1 g (B. Chen 1985; mean dimensions 3.50 × 5.97 cm, n = 178). C. Huang (1983) gave a range of values for egg measurements from the Shanghai Zoo (length: 56 mm–61 mm; width: 34 mm–38 mm; mass: 33 g–48.7 g). Based on a sample of 14 nests at the ARCCAR and 10 wild nests, we found mean egg size to be 5.71 (± 0.27) cm (egg length) × 3.46 (± 0.11) cm (egg width) and 41.4 (± 3.1) g. The largest individual egg we measured was 6.08 × 3.83 cm and weighed 53 g (from a captive female at the ARCCAR). The smallest egg measured was from a 2002 wild nest at Hongxin and it was 5.23 × 3.27 cm and weighed 32 g.

Even though egg size is generally correlated with the size of adult females, the eggs of Chinese alligators are smaller even than those of other crocodilians with smaller adult body size. In a review of crocodilian reproductive characteristics, Thorbjarnarson (1996) reported the mean length of adult female Chinese alligators was 151 cm and mean egg mass was 48.2 g. The smallest crocodilian, *Paleosuchus palpebrosus*, with a mean female TL of 108 cm, lays eggs that average 68.6 g (Thorbjarnarson 1996). The three other crocodilians smaller than Chinese alligators also lay larger eggs: *P. trigonatus* (adult female TL = 125 cm, mean egg mass = 67.2 g), *Osteolaemus tetraspis* (adult female TL = 131 cm, mean egg mass = 55.0 g), *Caiman crocodilus* (adult female TL = 143, mean egg mass = 62.9 g).

Clutch Size

Chinese alligators produce relatively small clutches of 20 to 30 eggs, although in proportion to body size, clutch size is larger than for many other crocodilians (Thorbjarnarson 1996). Captive alligators tend to have larger clutches than wild alligators (and may also nest at more frequent intervals), likely reflecting reduced energetic constraints and possibly larger female size. B. Chen (1985) reported mean clutch size of Chinese alligators was 20, with a range of 7 to 52. C. Huang (1983) gives a range of 8 to 41 for the Shanghai Zoo. Mean clutch size at the ARCCAR (1983–2005) was 27.2 (n = 1,171 nests). The largest reported ARCCAR clutch was one with 49 eggs (J. Xu et al. 1990). Similar clutch sizes are reported from the Changxing breeding center, where mean value from 1984 to 1991 was 23.1 (Webb and Vernon 1992) and one female (1.38 m TL, 9.8 kg) produced an average of 24 eggs during each of 4 years (W. Lu et al. 1988). The mean clutch size of 10 nests from this same breeding center was reported to be 23.6 by J. Wang and Huang (1997). The average clutch size had

Table 5.3   Wild nests and eggs recorded in the NCAR (1982–2006)

Year	No. of nests	No. of eggs collected	Mean eggs per nest
1982	10	224	22.4
1983	n.d.	n.d.	n.d.
1984	16	154	9.6
1985	4	80	20.0
1986	7	104	14.9
1987	5	99	19.8
1988	5	64	12.8
1989	1	19	19.0
1990	1	22	22.0
1991	n.d.	n.d.	n.d.
1992	1	23	23.0
1993	n.d.	n.d.	n.d.
1994	4	84	21.0
1995	2	41	20.5
1996	1	22	22.0
1997	3	60	20.0
1998	0	0	0
1999	2	33	16.5
2000	1	19	19.0
2001	0	0	0
2002	2	43	21.5
2003	2	49	24.5
2004	3	75	25.0
2005	4	108	27.0
2006	4	100	25.0
Total	78	1,423	18.2

Note: n.d. = no data available.

Table 5.4   Clutch characteristics of wild nests in the NCAR

Year	Site	Clutch size	Mean egg mass (g)	Clutch mass (g)
1999	Shaungken	17	42.5	722.5
1999	Zhongqiao	17	43.2	733.8
1999	Zhuangtou	18	39.6	712.8
2000	Zhongqiao	19	43.0	817.0
2002	Zhongqiao	18	39.7	714.6
2002	Hongxin	25	35.5	887.5
2003	Zhongqiao	22	42.8	940.5
2003	Hongxin	27	43.0	1,161.0
2004	Hongxin	28	44.7	1,251.6
2004	Hongxin	25	40.3	1,007.5

increased over the 2004 to 2005 period to 26.0. Wild nests (n = 78; 1982–2006) in the NCAR have a mean clutch size of 18.2 (Table 5.3), but if the abnormal 1984 data are excluded this figure increases to 20.5.

Eggs are not regularly weighed at the two main breeding centers, so estimates of clutch mass come entirely from wild nests we have examined during the course of this study (Table 5.4). Interestingly, clutch mass has shown a tendency to increase in wild nests since 1999, which may be, in part, a reflection of a greater number of recent nests at Hongxin where environmental conditions (e.g., food availability) may be superior.

Hatchling Size

One consequence of small egg size is that hatchling Chinese alligators are the smallest of the 23 species of crocodilians (Figure 5.25). Mean hatchling size has been reported to be 20 to 21 cm TL and 30 g (C. Huang 1982; B. Chen 1991). H. Zhu (1997) reports mean length and weight of neonates are 22.5 cm and 28.9 g. In 2003 we measured the size of hatchlings from nests at Zhongqiao and Hongxin; mean SVL, TL, and mass for hatchlings from these nests were 10.2 cm, 21.1 cm, and 27.5 g (Zhongqiao; n = 11), and 10.4 cm, 21.9 cm and 26.6 g (Hongxin; n = 22). Hatchlings from one of the two Hongxin nests in 2004 averaged 10.3 cm, 21.8 cm, and 26.3 g (n = 11). Based on the weight of the eggs measured at the beginning of the incubation period, neonates weighed between 58% and 64% of initial egg mass.

Nest Fate

A fundamental characteristic of crocodile life histories is a high rate of egg and juvenile mortality (Magnusson 1986; Abercrombie et al. 2001). Crocodilian eggs typically experience high rates of mortality as a result of unfavorable environmental conditions (e.g., flooding or excessive desiccation) and predation. Eggs in the nests of American alligators are taken by a variety of predators, mostly mammals such as raccoons (*Procyon lotor*) and black bears (*Ursus americanus*). There are few natural

Figure 5.25. Hatchling Chinese alligators are the smallest of all the crocodilians at birth, with a mean total length of 22.5 cm.

Table 5.5   Information on wild nests in Jinxiang County, Anhui (1984–1989)

Year	Site	Date found	Clutch size	No. hatched	% hatched	Date hatched
1984	Yan Tang	20 Jun 84	20	19	95	13 Sep 84
1985	Yan Tang	01 Jul 85	19	16	84	16 Sep 85
1985	Zhongqiao	27 Jun 85	23	18	78	16 Sep 85
1986	Yan Tang	29 Jun 86	34	26	76	18 Sep 86
1986	Zhongqiao	28 Jun 86	17	15	88	16 Sep 86
1987	Yan Tang	05 Jul 87	26	20	77	20 Sep 87
1987	Zhongqiao	29 Jun 87	22	20	91	16 Sep 87
1988	Yan Tang	03 Jul 88	28	15	54	10 Oct 88
1988	Zhongqiao	30 Jun 88	23	23	100	18 Sep 88
1989	Yan Tang	03 Jul 89	22	20	91	20 Sep 89
1989	Zhongqiao	28 Jun 89	18	18	100	14 Sep 89

Source: G. Wang et al. 2000.

predators remaining in the agricultural landscape of China, and we did not record any cases where alligator nests had been depredated. Likewise, the nature of the sites used for nesting has been so altered by human activities that nest flooding in most areas we saw was unlikely. B. Chen (1985) reported a 36% hatch rate for wild nests in Anhui but provided no details on the causes of egg mortality. Information on wild Chinese alligator nests from 1984 to 1989 in Jinxiang County in the NCAR was gathered by the county forestry officials, but no reasons for egg mortality were specified (Table 5.5). From a sample of 20 wild nests monitored from 1998 to 2006 we calculated an egg inviability rate of 13.6% (eggs did not develop bands), and a mean hatch rate of 62.3% (70.5% of the viable eggs). One nest (Zhuangtou 1999) experienced total egg mortality, which we attributed to low incubation temperature.

Nest Temperature Relations

Crocodilian eggs require a warm (28°C–34°C) and relatively stable thermal environment for successful embryonic development. The temperature at which eggs incubate determines a variety of important characteristics of alligators, the most apparent being gender. Temperature-dependent sex determination has been found in all species studied to date (Lang and Andrews 1994; Piña et al. 2003). For TSD species, the temperature of incubation also plays an important role in shaping other aspects of fitness such as hatching success, posthatching survivorship, and growth rates (Joanen et al. 1987; Lang and Andrews 1994; Janzen 1995; R. Shine 1999; Piña et al. 2003). In mound-nesting species such as the Chinese alligator, the nest mound serves as both a thermal insulator as well as a source of heat produced by the microbial decomposition of nest material (Ferguson 1985). Metabolic heat produced by embryos also contributes to elevated nest temperatures (Magnusson et al. 1985; Ewert and Nelson 2003).

The American alligator was one of the first crocodilians found to have TSD (Ferguson and Joanen 1983), but this early experimental work resulted in erroneous conclusions about the pattern of TSD and the sex ratios of hatchlings (Lang and Andrews 1994). The first studies of TSD in American alligators indicated that low incubation temperatures (≤ 31°C) produced females and higher temperatures resulted in the production of males. Experimental incubation of eggs of crocodiles found a different pattern, with females being produced at low temperatures and also very high temperatures (Hutton 1987) and this was thought to reflect the existence of two patterns of TSD in crocodilians. However, subsequent work has shown that alligators also produce high-temperature females, so while there is a relatively significant amount of variability among species in sex ratios produced at set temperatures, the overall pattern—males produced at intermediate temperatures—is similar (Lang and Andrews 1994). The thermosensitive period for American alligators, when sex is determined, coincides with the developmental differentiation of the gonads and is typically in the range of 30 to 45 days after oviposition (Lang and Andrews 1994).

The effects of incubation temperature on the sex of American alligators are relatively well known but the exact relationship is not yet clearly understood for Chinese alligators. TSD is known for this species (Lang and Andrews 1994), and staff at the ARCCAR indicate that the critical temperature (producing an even mixture of males and females) is 31°C (C. Wang, pers. comm.). B. Chen (1990b) reported that 91% males were produced at incubation temperatures above 33° to 35°C and temperatures below 28°C resulted mostly in females. Of 16 hatchlings produced from eggs incubated at 31°C by Herbert et al. (2002), 10 were male and 6 female.

Nearly all experimental work on TSD has been carried out in laboratories at constant or tightly regulated temperature conditions. Eggs in natural nests vary on a daily basis, and for most species nest temperature tends to increase during the incubation period (Ewert and Nelson 2003). This temperature variability may play an important role in sex determination (Webb et al. 1987), but this is poorly understood. We

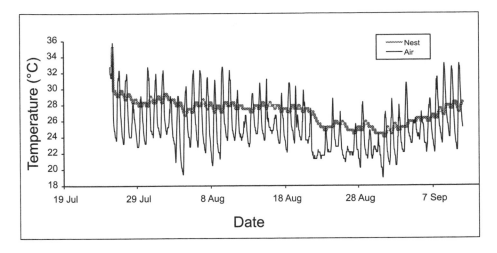

Figure 5.26. Air and nest chamber temperature of the 1999 nest at Zhangtou measured with a temperature data-logger.

were unable to carry out experiments on TSD, but we did measure nest temperatures in wild nests, and some under natural conditions in breeding centers. We measured the nest cavity temperature at a wild nest of Chinese alligators at Zhuangtou from 25 July through early October 1999 using a temperature data-logger (Figure 5.26). Nest and air temperature at three "natural" nests in the breeding enclosures at the ARCCAR were also measured with data-loggers in 2001. In 2003 and 2004 we used data-loggers to measure nest temperature in wild nests at the Hongxin site.

The first nest that we studied (Zhuangtou) was unusual. Temperature recordings indicated that for the first month incubation temperatures were low, but in the viable range. After mid-August, egg temperatures started to dip into the lethal range for embryos ($< 26°$ C) in association with a period of rain. One factor that may be important in determining the ability of nest mounds to produce heat via the microbial breakdown of plant matter is the type of nest material used and its suitability for microbial decomposition. Pine needles, the predominant material used in the Zhuangtou nest, decompose slowly (Gholz et al. 1985) and produce little heat. Furthermore, the Zhuangtou nest was fairly high on a ridge in a pine forest about 50 m above the ponds where the alligators lived, which themselves were in low hills. The relatively high and somewhat exposed location was reflected in the low air temperature (Table 5.6).

In 2003 we measured nest temperatures of wild nests at Hongxin (HX) and Zhongqiao (ZQ). Both these nests successfully hatched and exhibited a different nest temperature regime than the unsuccessful Zhuangtou nest. All nests showed a seasonal decline in temperature as mean air temperature fell in August and early September, and this appears to be a characteristic feature of Chinese alligator egg incubation. However, the successful nests had a much higher mean egg chamber temperature (ZQ = 30.3°C; HX = 29.0°C) (Table 5.6).

Mean nest temperature at the Zhuangtou nest was 1.8°C above mean air temperature. As this nest was shaded and did not receive any direct sunlight, this difference

Table 5.6   Air and nest temperature parameters for wild alligator nests

Year	Site	Mean	Min	Max	SD	Range	Mean daily range	Date range
1999	Zhuangtou							25 Jul–10 Sep
	Nest	27.09	24.00	35.20	1.50	11.20	0.56	
	Air	25.26	19.00	35.70	3.00	16.70	6.18	
2003	Zhongqiao							22 Jul–07 Sep
	Nest	30.31	25.90	32.70	1.62	6.80	0.43	
	Air	27.27	19.80	35.20	3.33	15.40	5.87	
	Hongxin							12 Jul–15 Sep
	Nest	28.96	21.30	35.20	2.64	13.90	2.04	
2004	Hongxin							23 Jul–31 Aug
	Nest	29.77	25.60	33.60	1.53	8.00	1.60	
	Air	27.00	19.00	35.70	3.46	16.70	6.42	

Note: Temperatures are in °C. All clutches successfully hatched except the 1999 Zhuangtou nest. The data-logger for the 2003 Hongxin air temperature did not function properly, so these data are not included.

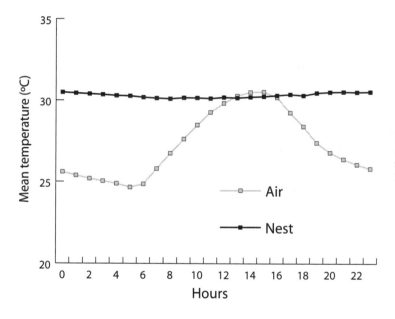

Figure 5.27. Mean diurnal air and nest temperature variation for the 2003 nest at Zhongqiao.

can be attributed primarily to heat produced in the nest mound, although metabolic heat of the embryos may have played a role. The Hongxin and Zhongqiao nests had a corresponding difference of 3.0°C and 2.8°C, respectively. The average daily temperature variation for the Zhongqiao nest is shown in Figure 5.27. These nests were made of a mixture of leaves and other parts of various broad-leaved vegetation and, in the case of the Zhongqiao nest, bamboo, materials that may be more susceptible to microbial breakdown. The effect of different types of nest materials on the ability of nest mounds to produce heat by composting is an intriguing mater for future investigation. Our work suggests that alligators find harsh conditions for nesting in the low hills surrounding the valley bottoms. By being forced up into hillside ponds like those at Zhuangtou, forced to nest in higher, cooler sites, and in some cases having to build mounds out of material that is more resistant to decomposition, alligators in these environments have a difficult time successfully incubating their eggs.

Another important ecological aspect of alligator nesting is the effect of rain. Periods of rain cause the nest mounds to get wet and lose heat. Because Chinese alligators make relatively small mounds and live in cooler environments than other crocodilians (with the exception of the American alligator), rainfall may play a more important role in their nesting ecology than it does for other mound-nesting species. By lowering nest temperatures, the pattern and frequency of rain play an important and relatively stochastic role in the resulting sex of the offspring. Moreover, extended periods of rain and its associated cool weather may result in increased egg mortality.

DIET

Alligators, like most crocodilians, are known to be very catholic in their dietary habits. Diet can be affected by a large number of variables, including the size of the alligator, as an ontogentic shift from small invertebrates to larger vertebrates is a general characteristic of crocodilian feeding habits (Webb, Manolis, et al. 1983; Fitzgerald 1988; Thorbjarnarson 1993).

Extensive dietary studies have been carried out on the American alligator (Delany et al. 1999).Other than a few early reports, however, very little is known about the natural diet of Chinese alligators. Hsiao (1934) noted stomach contents including remains of rats, beetles, water bugs, and fishes, and while captive animals would not eat puppies they did consume frogs, ducklings, and fish when offered. B. Chen (1985) analyzed the stomach contents of 14 wild-caught alligators, finding that the largest part of the diet was comprised of river snails (*Cipangopaludina* sp.; Pomacea; 41% by weight), spiral-shelled snails (*Bellamya* sp.; 22%), rabbits (*Lepus sinensis*; 16%), freshwater mussels (8.3%), freshwater shrimp (including *Caridina* sp., *Neocaridina* sp., *Palaemon* sp., and *Macrobrachium* sp.; 4.1%), frogs (2.6%), fish and Odonate larvae (2.3% each), and insect remains (1.9%).

The tendency for alligators to eat dead rodents poisoned by local homeowners and then tossed into their ponds apparently led to the deaths of two adult females during the 1990s (Anhui Forestry staff reported that both were found with a stomach full of mice). Hsiao (1934) reported that local people in the vicinity of Wuhu believed that small turtles formed a large part of the alligator's diet, but the author did not give much credence to these views. He indicated that alligators would take ducks and chickens, but pointed out that "in nature the alligators may be responsible, to some extent, for the missing of domestic fowls but this misbehavior is made good by their destruction of the most common and pestiferous rats and the part they play as scavengers of dead animals."

RECONSTRUCTION—A YEAR IN THE LIFE OF THE EARTH DRAGON

The life of the Chinese alligator remains somewhat enigmatic—there is still much we do not know. From several sources we have clues as to the types of habitats used by this species in the early twentieth century, before its wetlands habitats were fragmented and deteriorated. Unfortunately, this information is all from one particular area, the floodplain marshes in the vicinity of Wuhu. The Chinese alligator has proven itself to be extremely adaptable and ecologically malleable by surviving in the areas where it still occurs, and there is every reason to believe that when they were formerly widespread and numerous they were similarly flexible in their habitat needs. An analogous situation can be seen today in the wide variety of areas where the American alligator thrives in the southeastern United States. But that era has passed for the Chinese alligator. The fragmented nature of our knowledge of this species, and the need to rely on information about other species, is, for us, unsatisfactory. Undoubtedly, we will continue to learn more and more about this species, particularly as a result of monitoring how they adapt to new habitats as part of future reintroduction programs (Chapter 8). However, one of our goals in writing this book is to synthesize what we know about Chinese alligators and present it in a form coherent to those interested in the ecology and conservation of this remarkable animal.

In this chapter we have tried to pull together the available, although necessarily incomplete, information on the ecology and behavior of the Chinese alligator, augmented with the results of studies of other crocodilians, particularly the American alligator. However, it is hard to read through this material and get a real feel for what the Chinese alligator really is. To remedy this, in this final section on the behavior and ecology of the Chinese alligator we paint a picture of a year in the life of the alligator, illustrated by the annual cycle of hibernating in the winter and reproducing in the summer. Some of what we present here falls into the category of speculation, and in this sense we feel like paleoecologists, trying to reconstruct the former lives of extinct animals based on fossil evidence. However, we have more material to work with, and hopefully, some of the surmises we make will be subject to future testing on populations of alligators living in natural habitats.

Life Begins Anew: The Spring

We begin our narrative in the spring. The alligators have been hidden in their underground burrows for several months to avoid the lethally cold subzero aboveground temperatures that come in December, January and February. By early April the days are longer, and periodic warm, sunny intervals start heating the ground. Rain becomes more frequent and the small ponds and other wetlands on the alluvial plains begin to fill with water. The spring rains also start the rivers rising, eventually spilling over their banks and completing the flooding of the marshy plain by the early summer. Emerging from their dens in fits and starts, alligators bask only in the middle of the day during the first warm periods of the spring. On overcast or cool days the alligators remain hidden in their burrows. As temperatures continue to rise they spend more and more time out of the burrow. Later in the year, during the heat of summer, alligators will use the burrows to keep cool during daylight hours and emerge principally at night. Juvenile alligators remain in the burrows longer into the spring than the adults. When they do emerge, the young of the year remain close to the female during the day but begin foraging in the newly flooded marsh at night.

As spring advances, alligators become more active, feed more frequently, and show signs of sexual activity. The hormonal and gonadal cycles of males and females are closely tied with ambient temperatures. After alligators emerge from hibernation, the size of male testes increases and circulating levels of testosterone rise. Similarly, in reproductive females estradiol levels increase and vitellogenesis begins in response to increasing temperatures. Both male and female alligators start bellowing in the spring, mostly from the shallow water edges of their ponds. The short, explosive sounds of bellowing alligators, which are frequently associated with rain, define male breeding territories and bring the two sexes together for courtship and mating. The large, dominant males in the area either establish breeding territories that the females enter or they move around in search of mates. These males actively exclude smaller males from the breeding groups by advertisement displays (bellowing, headslaps), visual displays of body size, and, in the last resort, fighting. Courtship begins in May, either in the main ponds or other recently flooded depressions in the marsh. When alligators are close enough to each other, visual signals, vocalizations, and tactile cues (e.g., snout rubbing, bubbling) precede coupling. Breeding dynamics vary depending on the nature of the habitat and alligator population densities. In some areas, such as large ponds or lakes, alligators form groups in which dominant males mate with multiple females and females also pair with multiple partners, resulting in offspring of mixed parentage. In other habitats males move long distances to visit females that remain dispersed. Not all adult females nest every year, and in areas of dense alligator populations, females may nest only every second or third year. This is particularly true for the smaller, younger females. Courtship and mating peak in June, and shortly afterward the females who will reproduce prepare for nesting.

Summer: Nesting in the Marshes

Some of the females who do not nest this year did so the year before, and if they were successful still have some of their offspring with them. Those that will nest move to areas of suitable habitat, in some cases these are secluded, heavily vegetated sites with dens, a burrow that the female uses for shelter, and possibly a small pond (alligator hole), created or maintained by the female's actions. The peak of nesting occurs about 1 month after the height of courtship and mating. During this time females ovulate, and some 2½ to 3 weeks later lay their eggs. The calcium needed for laying down the eggshells inside the oviducts is mobilized from the mother's bones. The peak of nesting tends to coincide with the annual peak in water levels (late June and early July), and the nests are made on small elevated sites in a generally flat landscape. Up to 2 weeks before actually laying the eggs, the female begins making nocturnal visits to potential nest locations and flattening down the recent growth of grass and herbs. With backwards, scraping movements of her legs, she begins accumulating this material, which will be shaped into a small mound. When the nest is finally ready, she deposits 20 to 30 eggs into a chamber she opens in the top of the nest mound with her hind legs. The eggs, about the size of those of a chicken but more symmetrical, emerge from her cloaca one at a time, covered with a slippery mucous that keeps them from being damaged as they drop into the egg cavity. After she deposits the last egg she covers them with material from the nest mound that, when finished, is typically 40 cm high and about 1 m in diameter.

Nests are easy to spot shortly after construction as the female has cleared an area, usually with a 3– to 5–m radius, around the mound. After laying her eggs the female remains close by and periodically visits the nest, occasionally adding new material to the mound. Within a few weeks the vegetation surrounding the nest regrows rapidly and nest mounds are much more difficult to see. Predators such as raccoon dogs (*Nyctereutes procyonoides*), foxes (*Vulpes* spp.), and wild boars (*Sus scrofa*) among others are on the prowl looking for alligator eggs. But the nests are spread out and usually surrounded by flooded marsh, making them difficult to find. Eggs are also vulnerable to flooding if heavy rains persist. Some mound-nesting crocodilians nest on floating mats of vegetation, greatly reducing the possibility of nest flooding. As one of the few high and dry substrates in the flooded marsh, the alligators' nest mounds may be used by a variety of other reptiles, including turtles, snakes, and lizards, as a site to deposit their own eggs. The female alligator plays a role in keeping small egg predators away from her nest. If a predator opens the nest and takes some of the eggs, the female rebuilds the mound, covering the remaining eggs.

Eggs normally take about 60 days to incubate, but this depends on the temperature within the nest cavity. Eggs hatch faster under warmer temperatures. The temperature inside the nest during the middle one-third of incubation is critical for determining the sex and other characteristics of the hatchlings. The principal sources of heat for the eggs are the atmosphere (air temperature), direct solar radiation, and the heat of decomposition of the nest material. In effect the nest mound acts as a compost heap; the heat produced from the microbial breakdown of this material warms the eggs. Toward the end of incubation, when the embryos are large, the heat from their own metabolism in the close-packed environment also help maintain a high, relatively stable temperature inside the egg cavity.

The female has a biological clock that tells her when the embryos are reaching full development. Toward the end of the incubation period she will walk around or climb atop the nest. Inside the eggs, full-term embryos respond to the vibrations of the approaching female transmitted through the ground and start calling from within the egg. Once the female hears these high-pitched grunts, she opens the nest to free the young. Without the assistance of the mother, it is likely that few, if any, of the hatchlings would be able to escape from the nest. Many of the young alligators begin the process of escaping from the egg on their own by piping, using a small "eggtooth" on the tip of the snout to cut through the inner eggshell membrane and poke their head out though the egg. During incubation, the growing embryo has been consuming oxygen and producing carbon dioxide. As this CO_2 diffuses out through the eggshell and meets the atmosphere in the egg cavity, it forms carbonic acid, slowly eroding the eggshell. By the end of incubation the eggshell is much thinner than when first laid, allowing more gases to diffuse through the eggshell as the embryo grows, and also making it easier for the hatchlings to escape from the egg. If a hatchling cannot free itself from the egg, the female alligator will assist by picking up the egg in her jaws and gently squeezing the egg between her tongue and palate.

As the eggs hatch, the mother collects hatchlings in her mouth and carries them down to the water. Once in the water she may leave them along the shoreline near the nest or transport them to nearby dense vegetation or the entrance to her burrow. She will make several trips before all the hatchlings are out of the opened nest. The young alligators are born with a built-in food supply, the remaining yolk from the egg that has been drawn into their abdomen. Nevertheless, given the opportunity the hatchlings begin feeding almost immediately, grabbing at anything around them that is small and moves. At first the "pod" of neonates is extremely cohesive, particularly during the day when they spend much of their time basking in the sun around the edge of the pond piled on top of one another or even resting on the back or the head of their mother. In the heat of the day they seek shade, but as the sun sets they begin to disperse short distances around the pond or into the surrounding marsh to forage on insects and other small invertebrates.

In habitats where females are able to nest near their overwintering burrow, the young remain with the female in the vicinity of the nest or burrow. They may be joined by a few older juveniles from previous years. During their time at the nest and while moving, the hatchlings maintain communication with the female, using a repertoire of soft grunting vocalizations. The small alligators are vulnerable to a range of predators, including a variety of large wading birds and aquatic mammals such as the otter (*Lutra lutra*). If a neonate is seized by a predator such as a gray heron (*Ardea cinerea*), it emits a high-pitched cry—a distress call—attracting the attention of the guarding female. The hatchling is not likely to be rescued from a large wading bird, but the call alerts other hatchlings to the presence of a dangerous predator.

Hatchlings emerge from the nests in late summer and have, at most, 2 months before temperatures drop and they stop feeding. During this period the neonates grow relatively little, the most important matter being to survive through the first winter. Growth of juveniles is slow, and it takes 8 to 10 years before alligators (males and females) become sexually mature. Once the young alligators attain a certain size, they are viewed more and more as potential competitors by the resident adults and experience agonistic encounters with these larger individuals. Thus, larger juvenile and subadult alligators are prodded into dispersing from their natal areas in search of new habitats. Fitting in may be particularly difficult for the young males, who would be much smaller than the old, dominant males. Breeding opportunities for these small males are few until they can grow large enough to stand their ground against other males.

During the summer nesting season, adult and subadult males remain in more open, deeper water habitats, feeding and growing. At this time they will range farther than any other time of the year. If alligators have dispersed into new areas or have been forced to move as a result of drought or floods, they spend time digging new burrows. Even long-term resident alligators devote time to enlarging burrows.

Life Slows Down: The Fall and Winter

Temperatures begin cooling in late September and rains become less frequent. With the change in season, the alligators become less active during the cool nights and spend more time in and around their burrows. Water levels in the marsh are falling, and as temperatures drop the alligators start spending nearly all of the night in their burrows, only emerging on sunny days to raise their body temperature by basking. Once ambient temperatures start to regularly drop below 20°C foraging and feeding become less common and eventually stop when alligators remain in a torpid state deep in their burrows.

Females who have successfully nested will have their surviving offspring staying close. Of the eggs laid, many will be lost to floods or predators, and high rates of neonate mortality are common. The surviving babies have stuck near their mother and remain with her in her den. In habitats where females have had to nest in areas away from the winter burrow, as the waters in the summer flooded marshes recede females lead their young, either overland or through still-flooded patches, toward the site where they will overwinter. Due to their smaller size the juveniles cool down faster than the adults and enter hibernation earlier. At this time of the year they will be mostly active during sunny days when they can bask and raise their body temperature but will do very little feeding. Before the winter arrives they will spend some of their time in the female's burrow digging their own chambers to avoid being accidentally crushed by the much larger adult alligator.

In the burrows, in the late fall and early winter, groundwater levels are still relatively high and the alligators make periodic use of the "ventilation" holes for breathing. As the winter progresses and ground temperatures and water levels drop, the alligators move deeper and deeper into the burrows to avoid the coldest temperatures. The depth of the dens is commensurate with the level of the water table in winter, keeping the lowermost portion of the tunnel flooded and the den moist.

The alligators remain hidden in their burrows in a lethargic state for some 3 to 4 months, awaiting the return of warmer aboveground temperatures. Once ground temperatures begin to rise and longer periods of sunlight filtering down through the burrow herald the spring, the alligators begin to stir anew. Soon afterward they will leave their burrows and begin another year of life in the marshes along the lower Yangtze River.

鼉

6

The Dragon's Demise

Why the Chinese Alligator Disappeared

During our surveys for Chinese alligators, we visited Heyi village, a small community in what had once been part of the floodplain of the Zhanghe River, a southern tributary of the Yangtze. This small village was not one of the 13 sites in the NCAR designated for Chinese alligators; nonetheless it still held a small group of alligators and there were a few animals living in the village pond. Walking past bunched groups of two-story farmhouses, we skirted the long, winding pond, formerly part of a stream that had drained these alluvial lowlands. The farmhouses were boxlike with white tile facades, red roofs, and concrete foundations that in some cases extended out into the water, requiring us to make short detours inland around them as we continued along the edge of the pond.

In the past, summer floods caused the river to rise and to inundate low-lying parts of the surrounding area, creating a rich complex of wetlands for alligators and other aquatic fauna. Things are very different now. The river is confined within the banks of a levee. Those sediments washed down from the denuded upriver landscape that do not reach the Yangtze remain trapped within the levees, slowly raising the level of the riverbed. The river's swirling muddy waters have been lifted above that of the surrounding countryside, awaiting the opportunity to break through the imprisoning levees and flood the surrounding landscape. Winding streams like the one at Heyi have been decoupled from their former link with the river and divided into a series of elongate ponds used for fishing, raising ducks, and washing clothes.

Heyi village was one of the areas where almost 20 years earlier a Sino-American team of scientists had visited to count the alligators and make observations on their behavior. We first visited Heyi in 1999 and again in 2003, to count alligators and talk with the farmers that lived along the edge of the pond. But just one month prior to our second visit, rains had elevated the river level and it had escaped the confines of the levees. Many of the villages in this area had been flooded for several weeks, and we saw the lingering effects, particularly on the mud and stick houses of the poorer farmers. At one point we passed several houses that had collapsed into heaps of rubble (Figure 6.1). Asking about *tu long*, we were assured by the farmers that one or two still lived in the pond, but they were seen only rarely, spending most of their time hidden in their burrows, and no nest had been found in many years. When we engaged some of the residents in more conversation about the alligators, a few

Figure 6.1. Heyi Village after the flood in 2003.

expressed the opinion that many of their woes were the result of the *tu long*. During the springtime the farmers would occasionally hear the alligators bellowing, which they believed was the *tu long* calling for rain, a clear manifestation of its perceived dragon ancestry in the minds of some rural people For the residents of Heyi village, the alligator's bellows had triggered the excessive rains, which in turn caused the levee to fail. As we walked back up the muddy trail to our vehicle we talked with our local forestry department hosts about the irony of the situation. The alligator, a victim of the loss of wetlands, was now being blamed for devastating floods of human origin. The surge of agriculture that had over the millennia transformed the lowlands into rice paddies had made floods more likely through deforestation and the loss of wetlands that act as natural sponges to absorb the annual spring rains. The rivers that once transported water down to the Yangtze are now raised aqueducts perched above the neighboring fields, held back by levees that have to be repeatedly elevated. When the levees break, as they inevitably do, the result is devastating floods of the transformed landscape, and great human suffering. Although alligators and farmers are fellow victims, the dragon's offspring are now blamed for the very situation that has brought them to the edge of extinction.

ALLIGATORS IN SPACE AND TIME: THE YANGTZE RIVER

The China of today is two very different places. Western China, the roof of the world, is mountainous and sparsely populated and still contains a remarkable array of large wildlife species (Harris 2005). The lowlands of Eastern China, on the other hand, are densely settled, and little remains of its native fauna. These two worlds are spanned by the Yangtze River, or Changjiang as it is referred to by the Chinese, one of the world's mightiest rivers. The Yangtze is a product of the Himalayan orogenies and the monsoonal pattern of precipitation in southern Asia. It begins in the Tanggula Mountains, high on the Tibetan plateau, first running south and cutting a series of spectacular gorges through the eastern Himalayas in Yunnan Province before turning east and skirting the southern edge of the Sichuan basin, one of China's most populous regions. As the river continues eastward, it slices through the Wu Mountains, creating the fabled Three Gorges. Up to this point the course of the river has been guided by the geological structure of uplifted Western China. Once it passes Yichang, the Yangtze pours out onto the broad plain of Eastern China and enters one of the world's most densely populated landscapes. Here, the character of the Yangtze changes conspicuously, slowing and meandering through what is now China's breadbasket, eventually reaching the coast of the East China Sea at Shanghai, a sprawling megalopolis and the largest city in the world's most populous country.

Along the lower section of the river in Eastern China the adjoining lowlands are dotted with an impressive array of lakes in what was once one of the world's most biologically diverse alluvial plains, covered at one time by a mixture of broad-leafed forest and wetlands. In depressions along the course of the river, enormous lakes such as Dongting Hu and Poyang Hu formed interior basins that filled with the annual overflow of the Yangtze's summer floods. One of these areas, in southern Hubei, was known as the province with "thousands of lakes," and the adjacent wetlands were called *Yunmongzhe*, or "the endless marshlands" (Zong and Chen 2000). Historical records as well as remains from archeological excavations indicate that in the lower Yangtze region once supported a diverse assemblage of temperate and subtropical megafauna including elephants, horses, elk, bears, Malayan tapirs, muntjac deer, rhinoceros, bison, gibbons, and even a small relative (*Ailuropoda microta*) of today's giant panda. These species gradually disappeared from Eastern China with the concomitant increase in human populations after the lower Yangtze became one of the world's first centers of rice cultivation. This long agricultural tradition, extending back some 8,000 years, and the ensuing conversion of lands to crop monocultures have resulted in the exclusion or extirpation of the region's large wildlife. Today, the Yangtze River basin holds about 12% of the world's population, and the wetlands that once filled much of the alluvial plain have been almost completely co-opted for agriculture. The history of the settling of this part of China has been the story of development of a great human civilization but the loss of a remarkable fauna.

Two of the last remaining wildlife species living in the lower Yangtze illustrate the immense challenges for the conservation of faunal remnants in this landscape. In the river itself, the endemic Chinese river dolphin, or *baiji* (*Lipotes vexillifer*), may still survive by the thinnest of margins (X. Zhang et al. 2003). The sole surviving member of its family (Lipotidae), the *baiji* was historically found along 1,700 km of the lower sections of the Yangtze River from the river's mouth near Shanghai to Yichang. Human modifications of the river and floodplain habitats used by the *baiji* have been the principal reason for its decline. Riverine habitat has been severely degraded by dredging, the damming of the Yangtze and its tributaries, and the drainage or isolation of floodplain lakes. *Baiji* have died as the result of entanglement in fishing gear, electrocution from electric fishing, collisions with vessels, and blasting for channel maintenance or harbor construction (Reeves et al. 2003). Today, the *baiji* is the world's most endangered cetacean. Surveys carried out in 2006 failed to find a single animal, suggesting that the species is on the edge of extinction or may have already disappeared (Turvey et al. 2007).

The Chinese alligator is the *baiji's* more terrestrial counterpart, and like the dolphin it has dwindled to a level of ecological insignificance. Throughout the middle and lower Yangtze, the alligator's wetlands habitat has been converted into agricultural and urban landscapes. As early as the 1930s Sowerby stated that the alligator "is doomed to extinction, its haunts slowly diminishing as more and more land is being [converted to farmland], while the local fishermen appear to be killing off the few specimens that remain" (Sowerby 1936). The fact that it has survived into the twenty-first century, albeit in very small numbers, is a testament to its resilience and, as Fauvel noted, its tenacious hold on life. In this chapter we discuss how the habitat and distribution of the alligator have been progressively diminished over the last 2,000-plus years in Eastern China.

In the Neolithic and early historical times Chinese alligators lived far to the north of their present distribution as a result of warmer environmental temperatures. The loss of the northernmost populations of alligators had as much to do with global climate shifts as it did with habitat loss. However, the disappearance of wetlands and related human population pressures have been the principal cause of the disappearance of alligators from the Hangzhou Bay drainage systems and their near total extirpation from the Yangtze. There are few historical accounts of alligators in the Yangtze in the period before extensive human settlement, but based on what we know about American alligators it is likely that alligators would have been relatively common in wetlands associated with riverine floodplains. One historical account of King Mu, ruler of the Zhou Dynasty (fl. 976–922 BC), who led an immense army against invading groups in southern China, reported that thousands of alligators were lining the banks of the Yangtze River, across which his troops were to be ferried (D. Wang 1988). In the Ming Dynasty (1368–1644) *T'o* were still reported to be "extremely numerous in lakes and rivers" (Fauvel 1879). Qing Dynasty (1644–1912) references to alligators suggest that they were widely distributed through most of the lower and middle Yangtze River basin (H. Wen 1995; Figure 6.2).

By the late nineteenth century, however, alligators were becoming rare and hard to locate. Despite extensive efforts to obtain alligators, Fauvel (1879) was able to collect only six specimens. The Buddhists considered it meritorious to purchase and release wildlife, and Fauvel believed that the religious value of releasing alligators in temple pools was enhanced by their relative rarity. According to Fauvel, Father Pierre-Marie Heude, a French missionary-naturalist who studied turtles and spent 34 years (1868–1902) living and working in Anhui, Jiangsu, and neighboring provinces, had heard many times of the existence of local "crocodiles" but had never come across one himself. A visitor to Father Heude's museum in the late nineteenth century reported seeing one alligator specimen that had been captured near Wuhu (Kahler 1895), presumably after Fauvel's scientific description of the species resulted in heightened interest in scientific circles.

Nevertheless, when Fauvel was describing the alligator, small numbers of them could still be found along the lower Yangtze. Fauvel (1879) cited a newspaper report (*Shanghai Courier*, 11 March 1879) of an alligator found buried in the mud near Chinkiang (= Zhenjiang, Jiangsu Province), at the confluence of the Grand Canal and the Yangtze River. In Reynolds's account of finding the alligator engraving on

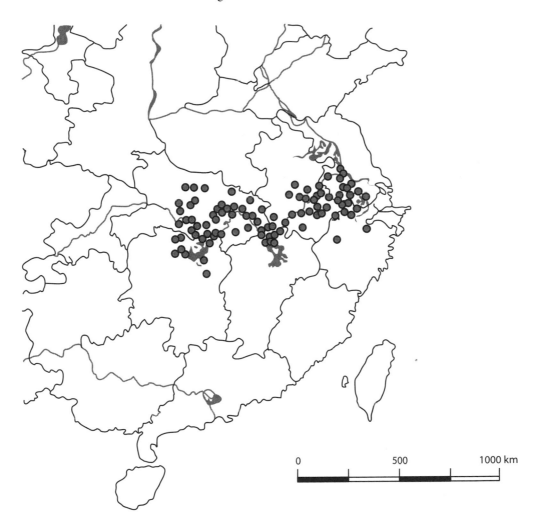

Figure 6.2. Historical locations of Chinese alligators during the Qing Dynasty. From H. Wen 1995.

Silver Island (Chapter 3), he noted that he had seen an alligator in the Yangtze River near the same island in 1876. When Fauvel himself visited the island, the priests told him that alligators were commonly caught in fishermen's nets and then usually killed and tossed back into the river. Fauvel was informed by others that these alligators were being washed downstream from the great lake systems (Dongting and Poyang Hu) by floods. Kahler (1895) mentioned reports that "lots of them" were dug out of burrows and brought to Shanghai in 1887 from a place referred to as Yangchow, also near Chinkiang. These alligators were reportedly sold to both Chinese and foreigners in Shanghai, the former releasing them into the Huangpu River (where most were subsequently killed) and the latter keeping them in miserable conditions in their bathtubs.

Throughout the twentieth century, the range of the Chinese alligator continued to shrink. In the 1930s the alligator was reportedly found "only in the great swampy lakes lying to the south of the Lower Yangtze," in Anhui, Jiangsu, and Zhejiang provinces

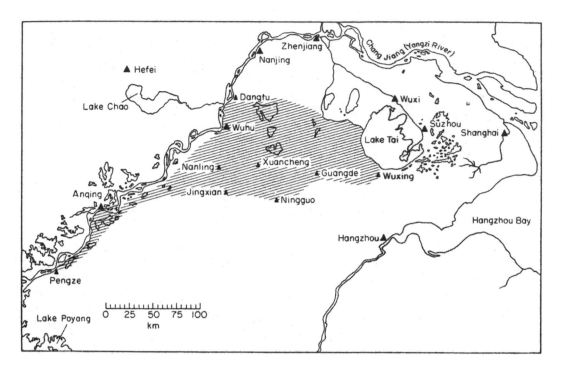

Figure 6.3. Reported locations of Chinese alligators in the 1950s. Modified from C. Huang 1982.

(Sowerby 1936). Their last stronghold was apparently the marshy lowlands to the south of the Yangtze River in Anhui Province (Chapter 3). A Roman Catholic missionary, Father Seckinger, told Fauvel (1879) that alligators were still "quite common" in his district (around Wuhu). He reported that "children play with the younger ones and larger specimens are captured in a long and narrow box. They are then killed and the skin, with head attached, is sold for medicine. The local name for them is *Tu Lung* i.e.[,] *Earth Dragon*." It was from this area that most animals were collected for museums and zoos in the late nineteenth and early twentieth centuries. In the 1920s and 1930s the range of the alligator was reported to be the lower Yangtze region. Small numbers of alligators were present in the vast Tai Hu lake in the 1930s (Pope 1935). Sowerby (1936) also reported that three individuals from Tai Hu were brought to the Shanghai Museum in July 1936. Nevertheless, during this period alligators were known principally from three "colonies" near Wuhu, all within a few miles of one another in seasonally flooded marshes along the fringes of Yangtze tributary rivers (Pope, in Schmidt 1927). The one colony that was visited by Pope was said to have "numerous" alligators, with 19 being dug out of their winter burrows in a period of less than a week.

As a result of continued killing and agricultural expansion, by the 1950s only small groups of alligators remained along the southern bank of the Yangtze River, extending from the vicinity of Pengze (Jiangxi Province) to the western shore of the Tai Hu lake and south to the foothills of the mountainous regions of southern Anhui (Huangshan Mountains) (Z. Huang 1981; B. Chen 1990a), and adjacent sections of Jiangsu and Zhejiang provinces (Figure 6.3). At that time alligators were reported to be living in lakes, some streams and marshes in lowland regions along the southern bank of the Yangtze, and in small ponds in low hills (Z. Huang 1981).

By the 1970s the range of the Chinese alligators was restricted to a small region in southern Anhui and neighboring Zhejiang Province. Z. Huang (1981) notes that during this time alligators were found less frequently in the typical habitat of low-elevation marshy wetlands, which had been almost completely transformed into rice paddies, but were increasingly restricted to small ponds in hills where agriculture and human disturbance were less intense. In some areas, local factors favored the survival of small groups of alligators in pockets of habitat within these agricultural landscapes, and in a few of these areas nesting continued.

In Zhejiang Province as recently as 1976, 3 alligators (2 females, 1 male) were captured in Huzhou County, Zhejiang Province, and sent to the Ningbo Zoo (M. Huang et al. 1987), and in September 1983, local residents in Anji County collected 18 eggs from a wild nest (M. Huang et al. 1987). In 1998 a single individual was captured in Anji County, Zhejiang, and brought to the Changxing breeding center. No other reports of alligators have come from Zhejiang, and our surveys suggest that only a few isolated individuals may remain (Chapter 7).

In Jiangsu Province there were reports of alligators in Yixing County in the 1960s, but a survey of this area in the early 1980s by Zhou Shie of Nanjing University failed to find any evidence of alligators (Ding 2001). There was one unconfirmed report of a farmer finding a juvenile alligator in 1994. In all likelihood alligators were extirpated from Jiangsu Province prior to the 1980s.

In Anhui Province alligators have persisted longer, probably as a result of the more recent initiation of agriculture in some marginal areas, a lower overall population density, and the greater prevalence of low hills with small irrigation ponds that offer a last refuge for alligators. The number of alligators in Anhui declined sharply between the 1950s and the 1970s. Of 29 sites where alligators had been relatively common in the 1950s, by 1976 they were easily seen in only 4 and had disappeared completely from 9 (B. Chen, in J. Fu 1994). Z. Huang (1981) reported that between 1956 and 1976, the number of known alligators in 11 counties in Anhui and neighboring provinces had declined from 360 to 120, an average annual decline of 3.3%. Understanding the patterns of alligator population declines since 1980 was one of the objectives of the field surveys we conducted from 1997 to 2005 and is presented in more detail in Chapter 7.

THE LEGACY OF YU THE GREAT: LOSS OF WETLANDS

The disappearance of China's wetlands is a story that began with the settlement of China's first Neolithic farming villages. Global climate shifts during the Pleistocene-Holocene transition played a significant role in the development of agriculture in several regions of the world, including China, approximately 10,000 years ago (T. Lu 1999). While historical references to rice and associated irrigation projects date back some 5,000 years, recent archeological evidence suggests that rice was domesticated in the middle Yangtze River valley much earlier, some 10,000 years ago (Z. Zhao 1998). However, while the Yangtze Basin was where rice cultivation first emerged, the developing center of the early Chinese civilization was farther north, on the dry North China Plain. In the late Neolithic period, agriculture based primarily on the cultivation of millet, and to a lesser extent rice (from about 6,000 BC), led to the development of urban centers in the Yellow River basin. From these Neolithic cultures were forged the earliest unified Chinese empire, symbolized by the forced imposition of a common system of writing during the Qin Dynasty in the third century BC and the development of a state ideology based on Confucianism in the second century BC. The early history of China is the story of a great economic power based to a large extent on its agricultural prowess, which,, in turn, was highly dependent on the availability and distribution of water. The demise of the Chinese alligator closely parallels the loss of wetlands in Eastern China, the product of China's population increase and the need for arable land.

The earliest written records of ancient China, from the Shang Dynasty (1700–1100 BC), consist of inscriptions in bronze or on oracle bones, the inscribed plastra of turtles or cattle scapula that were used for divination (Chapter 4). Records of Chinese civilization prior to the Shang are based on historical accounts written centuries later, in the Han Dynasty, most famously by Sima Qian. The first Chinese historiographer, he recorded legendary events from more than 1,000 years earlier in his work the *Shiji*, which was finished in 91 BC. One of the stories from the dim recesses of China's legendary past concerns a figure of particular note on the topics of wetlands—Yu the Great. The celebrated Yu, one of the most revered Chinese emperors, is considered to be the founder of the Xia Dynasty (2200–1750 BC). Yu ascended to the throne based on his accomplishments as a hydrological engineer after taming the floods of the unruly Yellow River. His father, Gun, had originally been given the task, which he set about by attempting to block the places where the river overflowed. After 9 years this approach clearly was not working, so a frustrated King Shun gave the job to Yu. He developed a new way to dredge the river channel, and after an epic 13-year struggle he won the battle. King Shun was so impressed with Yu that he passed his throne on to him. Thus began the tradition of the emperor assuming responsibility for managing China's waterways and controlling floods. Prior to the reign of Yu, the title of emperor was passed to the person considered by the community to have the highest virtue, instead of from father to son. But Yu's son, Qi, who was considered to be the most able, succeeded him as emperor, thus beginning the dynastic tradition of a hereditary monarchy.

The legend of Yu the Great is a metaphor for China's millennia-long battles to control, develop, and manage their rivers and wetlands, a struggle that continues to this day. While China is arguably best known to the world for the construction of its Great Wall, in terms of ancient infrastructure the completion of the Grand Canal, a series of waterways linking Beijing with Hangzhou, was on a similar scale in terms of cost and ambition. Two modern infrastructure projects have extended the tradition of Yu the Great in China. The well-publicized Three Gorges Dam, the world's largest hydroelectric project, backed up the waters of the Yangtze some 600 km and displaced more than 1.2 million people. The even larger-scale efforts by the Chinese government to divert water from the Yangtze north into the Yellow River basin are now beginning as part of a 50-year project. When finished, it will consist of three channels to connect the Yangtze River with the Hai, Huai, and Yellow rivers farther north. Each channel will be over 1,000 km long, and together they will carry about 48 billion cubic meters of water every year to drought-stricken northern regions, an amount equivalent to the entire current outflow from the Yellow River.

FILLING THE WETLANDS, CUTTING THE FOREST

During the Han Dynasty, two thousand years ago, China's population and agricultural center was still in the North China Plain and the lower Yellow River, where dry-land farming predominated and the principal crops were millet, wheat, and barley. Wilderness regions to the south were used as areas of banishment (Elvin 2004). Following the collapse of the Han Dynasty, China endured a lengthy period of disunity and wars between contending dynasties. During this time Mongol "barbarians," nomadic groups that lived to the north, repeatedly invaded China. As a result, many Chinese settlers moved south into the Yangtze basin, a region with a warm climate and fertile soils where wet field farming of rice dominated (Perkins 1969). While there is ample archaeological evidence of agricultural communities living in the Yangtze prior to the Han Dynasty (Chapter 3), during this period the clearing of land and draining of wetlands for agriculture in the lower and middle Yangtze began in earnest. By the sixth century AD the lower Yangtze was becoming fairly well settled.

Gradually, what had formerly been a hinterland in the Chinese empire became the new agricultural and economic center of the realm. When China was reunited under the Sui Dynasty (589–618), the emperor Yang Guang, realizing the importance of the Yangtze region, began constructing the Grand Canal, which in its entirety stretched more than 1,800 km from Hangzhou north to Beijing. By linking a series of existing canals, engineers created a waterway that allowed grain to be transported from the Yangtze to the Yellow River and the northern population centers.

The agricultural development of the Yangtze River valley and its connection with the major cities to the north by canals fueled a commercial and industrial explosion. China's population had remained at about 50 million from AD 2 to 750. During the late Tang and Song dynasties the expansion of rice cultivation in central and south China resulted in a doubling of the population to 100 million by 1100 (Figure 6.4). Marco Polo, who visited China during this period, noted that "this river [the Yangtze] goes so far and through so many regions and there are so many cities on its banks that, to tell the truth, in the total volume and value of trade on it, it exceeds all the

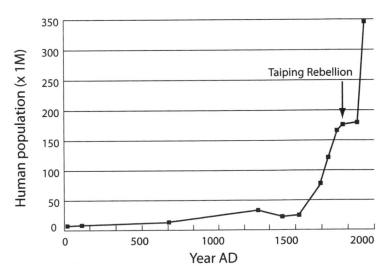

Figure 6.4. Human population of provinces in the historical range (Jiangsu, Zhejiang, Anhui, Jiangxi, Hubei, and Hunan provinces) of the Chinese alligator AD 2 to 2000. The graph clearly shows the effect of the Taiping Rebellion (1851–1864). There were approximately 20 million war-related deaths in China, and the overall impact (including other adult and infant mortality and declining birth rates) is estimated to be 40 million (Perkins 1969). Chart is from Durand 1960.

BOX 6.1

Rice Cultivation

Rice is a grass (*Oryza sativa*), originally from the wetlands of southern China and Southeast Asia, that has played a key role in the development of human civilization and culture. More than any other crop it has fueled the tremendous expansion of China's human population. Today, rice is the world's most important human food cereal. It thrives in low-lying, humid soils, is extremely adaptable, produces more calories per unit than any other cereal (1 ha supports 5.63 people, compared to 3.67 for wheat [T. Chang 1988]), and is the only major food crop that is almost exclusively a human food.

Neolithic peoples gathered wild rice to augment their fishing and hunting activities. Human use of rice evolved from shifting cultivation, a technique used for crops that depend on rains, to more specialized methods using naturally flooded areas and eventually to the construction of rice paddies where flooding can be controlled according to the seasonal rhythms of farming (Hanks 1992). Varieties of rice have been developed that allow more than one crop to be harvested in a growing season, effectively doubling the productivity of cultivated lands. Today, in most parts of the lower Yangtze, plots of land produce two annual crops of rice and an additional winter nonrice dry-field crop (usually wheat or rape seed).

In addition to being extraordinarily productive, rice is also much more suitable for long-term cultivation on plots of land. Depletion of soil nutrients and organic matter is a one of the principal factors that limits agricultural productivity. Rice, however, derives many of its nutrients from the water used in the annual flooding of the fields, as well as from the algae that blooms in the shallow, warm aquatic environments of the paddies. Rice is also a nearly ideal food. It can be easily stored and cooked, is highly digestible, tastes good, and provides good nutrition, particularly when consumed in conjunction with soy products.

Rice cultivation in China depends on a sequence of labor-intensive, well-controlled, and well-timed activities from plowing and preparing the fields, to planting seeds in dense beds, transplanting the seedlings into flooded fields, weeding, and (in modern times) applying fertilizers and pesticides. The crop takes about 100 days to mature. The fields are then drained and the rice harvested (usually) by hand, cutting the stalks with sickles. The stalks are threshed to winnow out the seeds, which are spread out on a hard flat surface to dry. They are then milled to remove the husks and polish the grains.

While it had long been a part of the Chinese diet, rice did not become a truly preeminent grain until sometime during the late Tang (618–907) and Song dynasties (907–1279). Lowland areas that had previously been too wet for cultivation were made accessible through technological improvements in damming techniques and the development of water pumps. These innovations allowed wetland marshes and land around lakes and rivers to be modified for cultivation, which was encouraged in many parts of China during the Song Dynasty through tax incentives (Ebrey 1996). From the late sixteenth to the early twentieth century (Qing Dynasty) there were large migrations of people from the north into the lowland Yangtze region (C. Huang 1982; B. Chen 1990a), and some of the last of the river's floodplain was converted to agricultural fields.

---

rivers of the Christians put together plus their seas." Later barbarian intrusions from the north in the eleventh and twelfth centuries following the collapse of the Northern Song Dynasty and the invasion of the Mongols to establish the Yuan Dynasty at the end of the Song Dynasty (late thirteenth century), accelerated the movement of people into the Yangtze River valley.

The direct impact of wetland loss on alligator populations can only be surmised from the historical records describing the construction of water control projects, the growth of land under cultivation, and the related effects of deforestation. As the population in the Yangtze region grew and the demand for rice and other crops increased, the amount of land under production was steadily expanded. Farmers also began employing more sophisticated cultivation techniques, such as irrigation, terracing, and the alteration of marshes, river margins, and shallow sections lakes by building dikes to create polders. Gradually, over the millennia, more and more land was converted to farming and an increasing diversity of water control structures filled the landscape. As demographic pressures increased so did demands on resources in the rivers and remaining lakes and streams with harvesting of fish, reeds, waterfowl, and other fauna. By the time of the Song Dynasty (960–1279), there were worries about the ecological stability of the lower Yangtze as a result of excessive land alteration (Osborne 1998).

The deforestation of the upstream watershed also resulted in increased flooding and drought throughout the Yangtze basin. Writing about wandering groups of farmers in the hills of southern Anhui Province, Mei Boyan (1786–1856) gave a memorable description of how forested hillsides held water and deforested slopes caused erosion, flooding, and the clogging of lowland fields with silt (Elvin 2004).

When I went to Xuancheng [in the East], in Anhui, and asked the country people about it, they all of them said that when the mountains had not been developed the soil had held firmly in place, and the stones had not budged. The covering of plants and trees had been thick and abundant, and after the rotting leaves had been heaped up for a few years they might reach a depth of two or three inches. Every day the rain passed from the trees to these leaves, and from these leaves into the soil and rocks. It passed through the cracks in the rocks, and drop by drop turned into springs. Downstream, the rivers flowed slowly. What was more, the water came down without bringing the soil along with it. When the rivers flowed slowly, the low-lying fields received their water without suffering from disaster. Furthermore, even after half a month without rain, the fields lying high up still obtained an influx of water.

These days, people have used their axes to deforest the mountains. They have employed hoes and plows to destroy the coherence of the soils. Even before a shower of rain has come to an end, the sands and gravels will be coming down in its wake. The swift currents fill up the depressions. The narrow gorges are full to the brim and cannot retain the mud-filled water, which does not stop until it reaches the lowest-lying fields, where it is then stagnant. These low-lying fields become completely filled, but the water does not continue to flow in the fields up in the mountains. This is opening sterile soils for farming, damaging the other fields where the cereals do grow, and profiting squatters who do not pay taxes, besides impoverishing registered households who do pay them.

Deforestation and the decoupling of the middle and lower Yangtze from much of its natural floodplain through the construction of more than 2,700 km of levees have resulted in disastrous floods, such as the one in 1998 that killed thousands of people and caused more than $20 billion in damage. And the frequency of major floods has been increasing. Historical records from the middle Yangtze tell of the increasing frequency of major floods since the Tang Dynasty, which is the result of deforestation and the loss of wetlands. In the Tang Dynasty (618–907) floods occurred an average of every 18 years. This increased to one flood every 6 years in the Song (960–1279) and Yuan (1271–1368) dynasties, and every 4 years in the Ming and Qing dynasties (1644–1911). In the 20 years from 1980 to 2000 there were 6 major floods, an average of 1 almost every three years (X. Liu et al. 2004).

One area with a detailed historical record of wetlands loss is the Dongting Lake complex. The lake, along the southern margin of the Yangtze in northern Hunan and Hubei provinces, is one of China's largest lakes, swelling every summer as it receives the overflow waters of the Yangtze. But the nature of the Dongting alluvial plain has changed dramatically over the last 2,000 years. During the Han Dynasty, Dongting was actually comprised of a series of marshes and smaller lakes fed by four rivers, southern tributaries of the Yangtze. To the north of the river there was a large lake, Yunmeng Lake, and associated wetlands (Yin et al. 2007). By the Tang Dynasty, the Lake Yunmeng had largely disappeared as a result of the uplifting of the Qinling-Dabie Mountains, creating hundreds of smaller lakes (Yin et al. 2007) in an area today referred to as the Jianghan Plain. As a result of the uplifting in the north and increased precipitation during the same period, the low-lying marshes to the south of the river were transformed into one huge body of water—Dongting Lake (Figure 6.5), in what is today Hunan Province.

During the Song Dynasty, the region around Dongting Lake was sparsely populated, with most settlements concentrated in the upland regions of the Xiang and Zi River basins (Perdue 1987). As pressure for farmland increased, more and more land was put under the plow. Efforts were begun to reduce the seasonal inundation of the floodplain and create farmland by building levees. Construction on the Jianghan Plain started as early as AD 346, and similar efforts along the southern margin of the river began between 454 and 464 (Yin et al. 2007). Efforts to alter parts of Dongting Lake were redoubled in the late Ming and early Qing dynasties as a result of increased water flow and sedimentation caused by changes in the lake's drainage pattern (Vermeer 1998) and by the filling of lakes in Hubei Province that had previously served as flood regulators (Perdue 1987). While there were still many medium-small lakes in the Jianghan Plain in the early Ming Dynasty, many of these had been silted up by the middle of the Ming Dynasty (1573–1620) (Yin et al. 2007). In the mid-1700s there were indications that wherever there was arable land in Hunan, farmers had cleared it, including deforesting hillsides and draining swamps (Perdue 1987). Some dikes were built by the government to control floodwaters, others, both sanctioned and illegal, were constructed around the lake by farmers (Vermeer 1998). In 1763 the region's governor reported that the lake was completely encircled by dikes that "spread like fish scales around the borders of the lake" (Perdue 1987). The reduction in natural wetlands caused floodwaters to rise, which meant the dikes had to be built still higher.

At the same time the effects of deforestation was causing significant amounts of silting in the ever more shallow lake. As a result of disastrous flooding events, the size of Dongting actually increased during the Ming and early Qing dynasties. West Dongting Lake was formed by the Yangtze River bursting through levees between 1570 and 1684, and South Dongting Lake was similarly created between 1852 and 1873 (Yin et al. 2007). Dongting reached its historical maximum size in 1825, when it measured about 6,300 km^2 (S. Zhao et al. 2005); it has subsequently diminished in size as a result of siltation and the conversion of shallow water areas to agriculture. Since the 1930s the size of the lake has decreased 49.2%, losing an average of 38 km^2 per year (S. Zhao et al. 2005; Figure 6.6). The degradation of the lake not only affects the region's fauna but also drastically reduces the lake's natural ability to retain floodwaters.

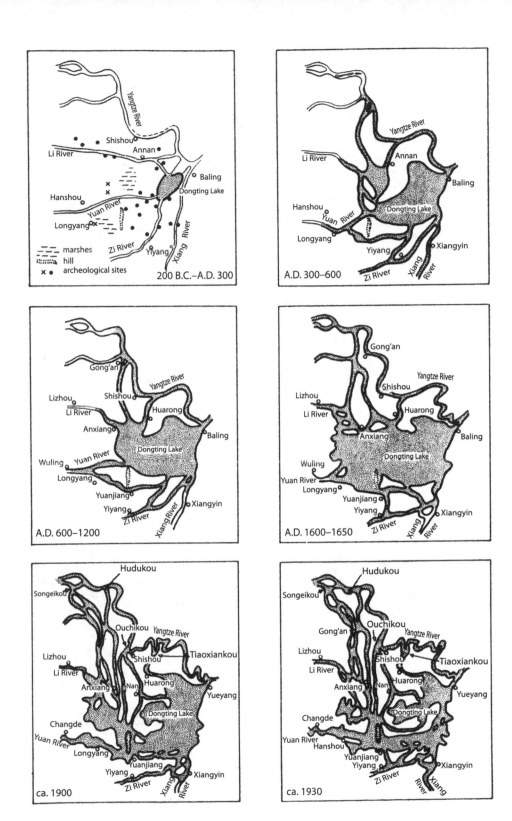

Figure 6.5. Historical evolution of Dongting Lake from 200 BC to about AD 1930.
From Perdue 1987.

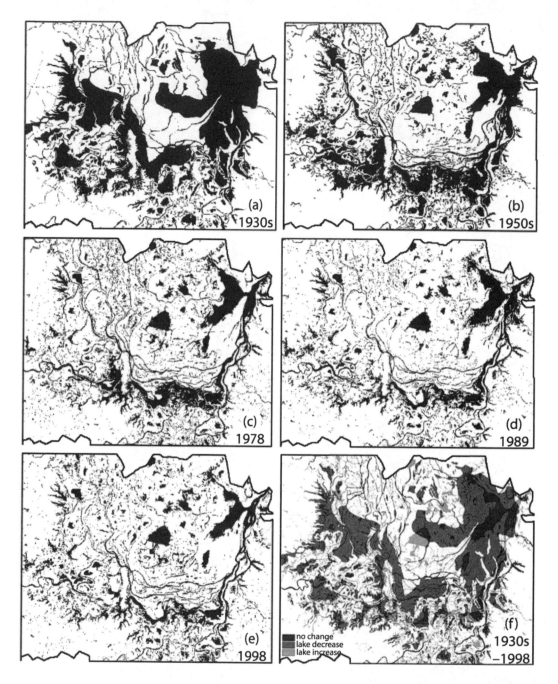

Figure 6.6. Loss of open water (shaded) area in Dongting Lake from the 1930s to 1998 is shown in (a) through (e). The loss (brown) and gain (gray) of open water are shown in (f). From S. Zhao et al. 2005.

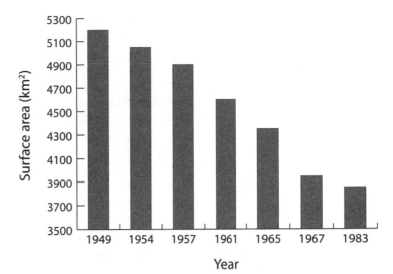

Figure 6.7. Surface area of Poyang Lake from 1949 to 1983. From X. Chen et al. 2001.

A similar set of events took place at Poyang Lake, another enormous lake along the south margin of the middle Yangtze. Once smaller than Dongting, Poyang Lake is now the largest in China, reaching a seasonal maximum size of over 4,000 km². Like Dongting, the history of building dikes around Poyang Lake extends back centuries and reached a peak in the late twentieth century. During Mao's "encircle lakes, create farmland" battle, the surface area of Poyang Hu was reduced by about 20% (Schapiro 2001). Before 1950 there were approximately 3,100 km of levees; that figure more than doubled over the next 50 years (Figure 6.7). At the same time, 1,300 km² of the lake were converted to land for farming and the surface area of the lake decreased from 5,160 km² to 3,860 km² (Shankman and Liang 2003).

The loss of wetlands has continued into modern times. The remaining arable lands have come under immense pressure from the expanding human population and the conversion of this region into China's breadbasket as well as one of its richest fisheries. Marsh has been turned into rice paddy; lakes have been isolated from the rising waters of the Yangtze River or modified for agriculture. Along the middle and lower Yangtze an estimated 1,000 lakes have been converted into farmland. Of 35,123 km² of lakes in the middle and lower Yangtze in the 1940s, 12,000 km² were drained for agriculture by the end of the 1970s (C. Fu et al. 2003). Poyang and Dongting lakes are representative of a widespread trend of lost lakes in the Yangtze River basin. Among a sample of 33 major lakes in the middle Yangtze River basin, only 2 were found to have increased in size between the 1950s and late 1990s, which was the result of extensive development of aquaculture projects along the lake perimeters (J. Fang et al. 2006).

Smaller lakes that have not been drained have been contaminated with industrial or agricultural runoff or are being used for aquaculture. Many of the larger water bodies that have retained their connections to the river, such as the Poyang and Dongting lakes, have become choked with silt, a process that has been exacerbated by rampant deforestation throughout the river basin. Siltation has increased to such a level that along parts of the middle Yangtze the riverbed is actually 2 m higher than the surrounding floodplain (Zong and Chen 2000).

One of the last areas where small groups of wild Chinese alligators survived was in northeastern Zhejiang Province, in the vicinity of Changxing (the site of one of the two operating breeding centers; Chapter 8). The destruction of wetlands in this region began along the southwestern shore of the Tai Lake, near the city of Huzhou, in the early eighth century, and then extended into surrounding areas. Records from 1377 indicate that by that time extensive land alteration had occurred as the land coverage in the region was 51% rice paddy, 34% hilly terrain, 11% dry agricultural fields, and 4% marsh (Yoshinobu 1998). Along the Ning-shao Plain in the Hangzhou Bay region, which historically had alligators, there were 199 lakes at the beginning of the Ming Dynasty; only 44 remained by the twentieth century and these had been greatly reduced in size and water storage capacity (Osborne 1998).

In the Yangtze River delta the building of extensive seawalls in the eighth century marked a period of extensive land modification. The marshy regions around the Tai Lake were the major focus of land modification and colonization, a process achieved principally by digging a network of canals to drain the land (Yoshinobu 1998). Between 1170 and 1225 new dikes and polders resulted in the draining of nearly all the seasonally flooded lands from Lake Tai to the sea in what is now Shanghai municipality and northern Zhejiang Province (Hartwell 1982).

Today, virtually the entire historic distribution of the Chinese alligator is under cultivation (Watanabe 1981). More and more, as the last natural habitats have disappeared, the alligators have been forced to occupy man-made water bodies. Many of the last areas where alligators are found today are small ponds created in the late 1950s and early 1960s during the period of the Great Leap Forward, with its emphasis on increasing rural industrial production.

Last Refuge—Southern Anhui and Adjacent Zhejiang

One question that remains unresolved is why southern Anhui and parts of adjacent Zhejiang Province remained as the last stronghold for Chinese alligators. By the twentieth century, the distribution of the Chinese alligator appears to have been reduced to the southern fringe of the Yangtze River from lake Tai Hu in the east to northern Jiangxi Province in the west (Figure 6.3). The populations of alligators in the Yangtze floodplain along the southern bank of the Yangtze disappeared during the course of the twentieth century, presumably as a result of human population pressures but also possibly as a result of winter floods in 1957. Such floods drown alligators in their burrows (Watanabe 1982). When the species was first discovered in the mid-nineteenth century, the area just to the south of the city of Wuhu, a port city on the Yangtze in southern Anhui Province, was particularly noted for alligators. This region was a broad alluvial plain with extensive marshland formed by an internal deltaic system of the Qingyi and Shuiyang rivers, which joined before flowing into the Yangtze. Each spring these plains flooded extensively and remained inundated for a good part of the year, impeding attempts at agriculture. Not until the early twentieth century were efforts made to control the flooding of this area (Box 6.2). Another factor may have been that at least part of the region was a de facto nature reserve during the Qing Dynasty, when the ruling mandarins set aside parcels of land

BOX 6.2

Wetlands Lost to Agriculture—The Case of Wanchun and Yitai Villages

B. Chen (1990a) provides an illuminating account of the conversion of wetlands to agriculture that occurred in the twentieth century in two villages, Wanchun and Yitai, located on the southern outskirts of the city of Wuhu, and how this impacted the wild populations of alligators. To the east of the city of Wuhu, the Shuiyang River flows westward, meeting the Qingyi; they then empty into the Yangtze just southwest of Wuhu. Both of these rivers collect water from a large area that now includes the NCAR. The region where the two rivers met, a low-lying, marshy area along the southern bank of the Yangtze, was one of the last strongholds of the species in the late 1800s and early 1900s. Reports suggest that alligators were found everywhere in this floodplain at the turn of the century. Residents of this region remembered that "alligators could be heard roaring everywhere in summer, as much as frogs are heard croaking."

The beginning of the end for this alligator population came in 1904 when work began on the Wanchun embankment. This structure, a levee along the major river course, eliminated the annual flooding of much of the surrounding marshlands, thereby opening the region to further agricultural development. Gradually, over the next five decades, the natural marshlands were converted into agricultural fields. The levee construction around the junction of the Shuiyang and Qingyi rivers was finished in 1931, creating the Yitai embankment. Studies in the 1930s (Hsiao 1934) still reported good numbers of alligators in ponds and water canals in the area. But alligators, which were widely viewed as a threat to agriculture and fishing, were killed whenever they were found. Alligators were known to occasionally eat domestic ducks and were believed to reduce fish populations. Their propensity to burrow into water control dikes made alligators unwelcome residents of ponds around villages and rice paddies. By the 1950s alligators were rare but still present, and they formed the basis of a study carried out in 1951, 1954, and 1956 (Chu 1957). In 1956 a final water control structure, the Wanchun floodgate, was built, converting the entire region into agricultural fields. During construction, alligators were dug out of their dens and killed. A few alligators survived, but a campaign to eradicate snails and schistosomaiasis was begun in 1958. The widespread application of the pesticide sodium pentachlorophenate resulted in the death of the last individuals in the early 1960s.

Alligators have also been the victims of drought and flooding that has resulted from the extensive deforestation and loss of wetlands in Eastern China. Many alligators probably drowned in their burrows along the Yangtze during floods in the winter of 1957 (Watanabe 1981). Flood and drought also may force alligators to move overland where they are captured or killed (B. Chen 1990a). Y. Zhou (1997) reported the mortality of 80 alligators in the Shuming village (Langxi County) due to drought in 1987.

The last individuals at Liangzhong in Jinxiang County reportedly left the area during the drought of 1997. At Hongshugang (Langxi County) alligators were regularly seen until the floods of 1983, when many were swept into the surrounding floodplain or washed downstream in the Langchuan River. In some areas farmers blame alligators for floods, claiming that the nighttime bellowing of alligators in the late spring and early summer brings the rains that burst the river levees.

for cattle ranching and prohibited farming, which effectively delayed the onset of agriculture and resulting loss of wetlands. One of these areas was the natural marshland in the Yangtze floodplain to the south of Wuhu (B. Chen 1990a). With the loss of these habitats some of the surviving alligators may have simply moved up Shuiyang and Qingyi rivers into the region that is now the National Chinese Alligator Reserve and the last area to harbor wild alligators. Agriculture in this region was not initiated intensively until the 1950s and 1960s as much of the area was marshy and tended to flood. Only after the relatively recent installation of adequate water control structures did large numbers of people (mostly from northern Anhui Province) move into the region, something that is reflected in the fact that human population densities in this area are much lower than in northern Anhui Province (especially

Table 6.1   **Human population density in the NCAR and percentage increase (1935–2005)**

County	2005 population density (per km²)	% increase since 1935
Nanling	435.1	220.2
Xuancheng	390.8	203.6
Langxi	307.7	247.0
Guangde	235.6	278.3
Jinxiang	174.8	165.0

the area just north of the Yangtze). Nevertheless, the number of people in the region has increased dramatically since the 1930s (Table 6.1).

Other accidents of human history may have played a role in preserving the last groups of alligators in this region. During the mid-nineteenth century, southern China was the site of one of mankind's bloodiest conflicts, the Taiping Rebellion. This conflict resulted in the deaths of some 20 to 30 million Chinese. The central and lower Yangtze regions were the hardest hit, and much of the region was effectively depopulated (Figure 6.4). As a result of this conflict, the population of four provinces (Anhui, Zhejiang, Jiangxi, and Hubei) was lower in the 1950s than it had been before the war (Perdue 1987). The effects of this conflict and the resulting changes in human occupancy and land use on the status of surviving wildlife populations can only be surmised.

FACTORS RELATED TO HABITAT LOSS

Viewed over the long span of Chinese history, habitat loss has been by far the most significant factor leading to the current endangered status of the Chinese alligator. However, a number of other indirect as well as more direct human-related causes have contributed to the overall decline and local extirpation of alligators in the Yangtze basin.

Direct Killing of Alligators

Hunting is a primordial human activity and early humans in eastern Asia were very generalized predators, consuming a wide variety of prey. With the growth of human civilization, hunting began to take on a new meaning for the social elite. The great royal hunts of the Shang Dynasty in the second millennium BC would target a wide variety of species, including tigers, elephants, and rhinoceroses, merely for sport. The Chinese were also among the first civilizations to put restrictions on hunting. The Ordinances for the Months, written during the first millennium BC, prohibited the use of particular hunting and fishing methods during critical times of the year, principally as a guard against overexploitation of species used for food (Elvin 2004).

Alligators have long been one of the species targeted for killing by people, either for use as food or because they were viewed as pests that interfered with the complex water management systems in rice-growing areas. Alligator remains have been found at sites used by *Homo erectus* in the middle Pleistocene (Bakken 1997) as well as by modern *H. sapiens* at Neolithic sites (Chapter 3). Archaeological studies have found that Chinese Neolithic cultures fashioned ceremonial drums from alligator skins (Chapter 3). The first historical reference to alligators being killed as pests was in the Ming Dynasty (1368–1644) when the banks of the Yangtze River near Nanjing were reported to frequently collapse due to the number of alligator burrows; large numbers of alligators were trapped and killed. During the early Ming Dynasty, because the name of the alligator was the same as the surname of the

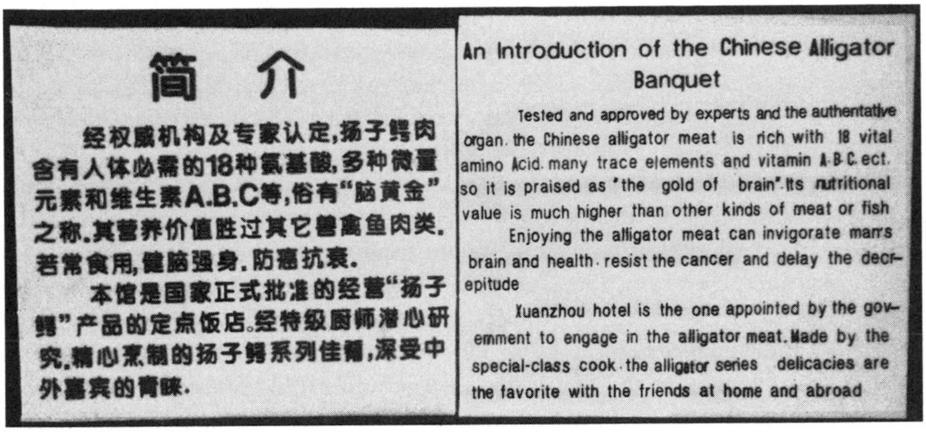

简 介

经权威机构及专家认定,扬子鳄肉含有人体必需的18种氨基酸,多种微量元素和维生素A.B.C等,俗有"脑黄金"之称.其营养价值胜过其它兽禽鱼肉类.若常食用,健脑强身.防癌抗衰.

本馆是国家正式批准的经营"扬子鳄"产品的定点饭店.经特级厨师潜心研究,精心烹制的扬子鳄系列佳肴,深受中外嘉宾的青睐.

An Introduction of the Chinese Alligator Banquet

Tested and approved by experts and the authentative organ. the Chinese alligator meat is rich with 18 vital amino Acid. many trace elements and vitamin A.B.C. ect. so it is praised as "the gold of brain". Its nutritional value is much higher than other kinds of meat or fish

Enjoying the alligator meat can invigorate marrs brain and health. resist the cancer and delay the decrepitude

Xuanzhou hotel is the one appointed by the govemment to engage in the alligator meat. Made by the special-class cook. the alligator series delicacies are the favorite with the friends at home and abroad

Figure 6.8. Sign in a restaurant in Xuancheng extolling the benefits of eating Chinese alligator meat. The restaurant is one of a small number authorized by the government to sell meat from the alligator breeding centers.

contemporaneous emperor, he ordered the people to catch and kill the giant softshell turtles (*Rafetus swinhoei*) instead of alligators (B. Chen 1985). Apparently the edict did not halt the capture and killing of alligators as other accounts mention digging them out of their burrows, using traps as they emerged from the burrows, or baiting hooks with roast dog. As a result, alligators were rare around Nanjing by the middle Ming Dynasty (B. Chen 1985). The classical Chinese Materia Medica, written during the Ming Dynasty, states that their meat was considered a delicacy and a favorite for wedding feasts (Fauvel 1879; B. Chen 1990a). Alligator scales were also used in treating syphilis (Gordon 1884).

Today, alligators are rarely tolerated by local farmers for a variety of reasons. Alligators prey on small livestock (particularly ducks), and their burrowing makes holes in water control dikes. Many farmers also believe alligators reduce the number of fish in ponds. In the mid-twentieth century, farmers considered alligators to be vermin and routinely killed them by placing hooks, baited with turtles, at the entrances to their burrows (Y. Wang 1962). Deliberate killing of alligators by farmers was still fairly common in the early 1980s (Watanabe 1981). There is some indication that the tradition of eating alligators continued, at least in some areas, into the twentieth century. Hsiao (1934) states that when "convenient" alligators are killed by farmers for food. Y. Wang (1962) indicates that alligator meat was considered a delicacy. Other accounts suggest that during most of the nineteenth and twentieth centuries, people did not eat alligators. If an alligator was killed it was usually chopped up and fed to ducks (PRC 1992) or to pigs (Webb and Vernon 1992). Whatever traditions there may have been for eating alligator meat clearly disappeared as the animals became increasingly scarce. But with the surge of captive breeding, some evidence suggests that in the mid-1980s people in Anhui Province began eating alligator meat because of claims that it was "dragon meat" and provided health benefits (B. Chen, pers. comm.). While the sale of alligator meat from the ARCCAR has been approved at a few restaurants (in Xuancheng [Figure 6.8], Hefei, Huangshan, and Beijing), an illegal market has also emerged. In recent years there have been instances of people catching and selling live alligators. In early August 1997 a man in Yuhang County, Zhejiang Province, was jailed after being found with a 1.6-m TL alligator that he had purchased in Anhui Province. In 1999 four men were jailed in Jinxiang County for the capture and attempted sale of a large male alligator (1.7 m TL). The men were seeking buyers in Guangzhou Province.

In the late twentieth century the capture of live alligators for zoos or breeding centers also played a significant role in the demise of wild populations. During the 1950s many young alligators were collected from Zhejiang and Jiangsu provinces under the direction of the Shanghai Zoo. Many of these animals were reportedly sent to the former USSR (B. Chen, pers. comm.), presumably to supply zoos. Some of the last individuals in Zhejiang Province were collected in the 1970s for zoos and in the 1980s to establish the Yinjia village alligator breeding center. In Anhui Province in the early 1980s, a total of 212 animals were collected to stock the ARCCAR. This wave of captures was the last straw for alligators in some valley bottom sites and probably had serious consequences for the survival of certain hillside groups in Xuancheng, Nanling, and Jingxian counties. The collection of eggs from wild nests in the NCAR continued until 2002. We provide a more detailed analysis of the captures in Chapter 7.

Over the last century, as the size of the population of wild Chinese alligators has diminished, the accidental killing of alligators, while infrequent, has become an increasingly important source of mortality given the small size of the wild population. In the NCAR, alligators have been accidentally killed by fishermen and by hunters. Reported causes of alligator mortality are summarized in Table 6.2.

Pollution

The three greatest threats to China's vertebrate fauna are habitat loss, overexploitation, and pollution (Li and Wilcove 2005). Industrial pollution and the widespread use of agrochemicals has likely been a particularly relevant factor in the loss of much of China's wetlands fauna.

Table 6.2   Causes of alligator population declines

Site	Cause of mortality or emigration (year)
Shaungken	Loss of habitat from dike construction (1991)
Zhongqiao	Hatchlings dying during winter from lack of burrows (1989–1992)
Yantang	Accidental shooting of last adult female while rabbit hunting (1995)
Yantang	Lower water temperature, increased current, more variable water level from stream hydrology changes (1992)
Liangzhong	Pond drying up (1997)
Hongxin	Last adult female killed eating poisoned rodents (1999); egg collection (prior to 1996)
Zhuangtou	Egg collection (1996, 1997)
Zhucun	A juvenile accidentally killed by eel fishermen (1998)
Changle	Last adult female killed eating poisoned rodents (1998)
Xifeng	Dam broke; pond emptied (1991)
Heyi	Floods (1999)
Huang Shugang	Floods (1983)

Note: Data are for sites visited during 1999 survey.

Table 6.3   Pesticide labels from Anhui Province, China (1999)

Compound type	LD_{50}[a]	No. of sites[b]	Compound type	LD_{50}[a]	No. of sites[b]
Herbicides			Insecticides		6
Metsulfuron-methyl	>5,000	8	Methamidophos[d]	13	4
Acetochlor	2,953	8	Buprofezin[e]	2,200	2
Bensulfuron-methyl	>5,000	6	Cypermethrin[f]	247	2
Glyphosate	5,600	4	Phoxim[e]	1,845	2
Haloxyfop	518	2	Fenvalerate[f]	451	1
Benthiocarb	1,300	1	Methyl parathion[d]	9	1
Sofit N	6,099	1	Carbofuran[g]	8	1
Onecide-P	3,328	1	Dichlorvos[d]	25	1
			Isoprocarb[g]	178	1
Fungicides			Deltamethrin	128	1
Validomycin A[c]	n.d.	4	Imidocloprid[h]	450	1
Tricyclazole-sulfur	314	1	Monosultap[c]	n.d.	n.d.

Note: Unidentified labels not included;
        n.d. = no data available.
[a] Acute oral LD_{50} for laboratory rats.
[b] Frequency of occurrence at 18 sites visited.
[c] Could not locate any information on this product.
[d] Organic phosphate.
[e] Thiadiazine.
[f] Synthetic pyrethroid.
[g] Carbamate.
[h] Neonicotinoid.

Beginning in 1958 programs to eradicate schistosomaiasis, a human disease (liver flukes) whose intermediate hosts are snails, resulted in the application of large amounts of sodium pentachlorophenate to agricultural fields, which killed alligators (B. Chen 1990a). After 1949 large amounts of chemical fertilizers and insecticides were used in cultivated fields, reducing or in some cases virtually eliminating the prey base on which alligators depend (B. Chen 1990a). Near cities, pollution of waters with heavy metals is common (Scott 1989). In our surveys from 1999 to 2005 we found clear evidence that alligators were also dying from eating rodents that had ingested poison in nearby houses. Y. Zhou (1997) reported that in 1984 two alligators died from the effects of insecticides in a farm in Jinxiang County. The use of pesticides for crop protection, and to some extent for home pest control, is a significant feature of the socioeconomic structure in the study area. Rice fields dominate most of the lowland landscape, and most waterways designated as alligator habitat are either encompassed by or adjacent to significant tracts of rice fields. In most cases effluents from rice fields drain directly into alligator habitats.

This intensive agriculture activity in and around current areas designated as alligator habitat greatly increases the opportunity for direct and indirect effects from exposure to agricultural pesticides. Survey results from 1999 indicate intensive use of insecticides and herbicides in rice cultivation (Table 6.3). Some of the insecticides are known to be quite toxic to mammals and birds (methyl parathion and carbofuran), but these chemicals were observed at only one site each. Methamidophos was the most frequently observed insecticide across sites; like methyl parathion and carbofuran, it is relatively toxic to laboratory rats. The level of toxicity of any of these pesticides to alligators is difficult to determine. Most toxicity data in reptiles is for older chemicals such as organochlorines. Nevertheless, under the right conditions, exposure to these insecticides would undoubtedly result in some level of effect, whether lethal or sublethal.

Based on eyewitness accounts and interviews, we conclude that pesticide use occurs throughout the study area during the entire year (both summer and winter crop production), overlapping with the breeding season of alligators, including the occurrence of hatchlings. Given this scenario, potential risks of exposure could result in both direct and indirect effects on different age classes of alligators. Under normal agricultural practices, direct effects from exposure to pesticides would most likely result from consumption of contaminated prey. This exposure would probably be most significant in hatchling and yearling alligators that consume large amounts of small prey items (e.g., invertebrates). Larger alligators are probably at much lower risk because of their size (buffering), diet, and feeding behavior. Large alligators are more likely to consume large prey at less frequent intervals (decrease exposure potential). Pesticides may also affect alligators through indirect routes, including local depletion of food resources in alligator habitats because of pesticide runoff from rice fields.

Habitat and Population Status

The expanses of Yangtze floodplain wetlands that were once home to the Chinese alligator have disappeared. The seasonally flooded marshes to the south of the city of Wuhu, in Anhui Province, were the last known location where alligators were found in an extensive alluvial wetland, a habitat that we believe was characteristic for this species in the past (Chapter 6). That landscape, once a broad marshy plain, has been completely converted to agriculture, and the last alligator survivors were found in a few ponds into the 1970s and the early 1980s. As farming, with its associated habitat modifications, has marched up from the plains into the hill valleys, it has pushed the alligators before it. When the alluvial plains around Wuhu disappeared, the only suitable areas for alligators was in wetlands upstream along the Qingyi, Shuiyang, and Langchuan rivers and their tributaries that drain the Huangshan Mountains. When the valley bottoms along these drainages were cultivated, the last alligator refuges in the adjacent hills were ponds created by farmers who dammed small streams to store water for the seasonal flooding of the river paddies in the valleys below. Here the last alligators survive in man-made ponds, a very few of which are down in the rich valley bottoms, but most of which are in marginal habitats along the rice-forest ecotone or in ponds up in the hills where soils are only suitable for tea or tree farms. It is in this area that the National Chinese Alligator Reserve was established (Figure 7.1).

HABITAT: HOME FOR A FUGITIVE SPECIES

When we started our surveys in 1997, the NCAR covered an impressive sounding 433 km². Since that time, the overall size of the reserve has been reduced to 308 km²; still a large region that conjures up images of a protected area with suitable habitat where alligators, and other fauna, can thrive. The truth is quite different. The human-dominated landscape of the NCAR is in a five-county area with more than 2.75 million people, where the greenest areas are agricultural fields. Nowhere are there natural wetlands. When we conducted our initial surveys the only locations that could be even remotely considered part of an actual "reserve" were 13 small sites that had been designed for the protection of the alligators and that altogether totaled 41 ha. These areas were tiny, as well as marginal in terms of habitat quality, and the ponds were

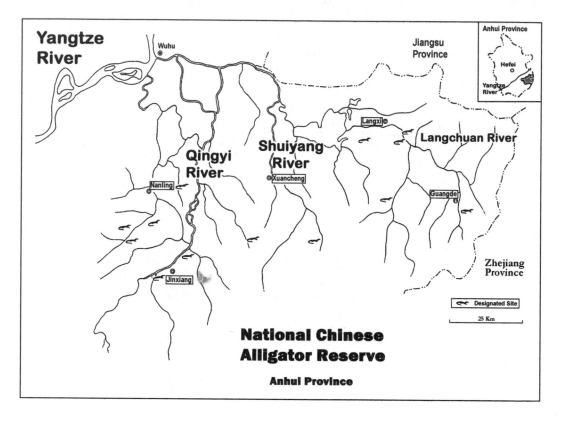

Figure 7.1. Location of the main NCAR sites with Chinese alligators.

not managed in any way by the Anhui Forestry Bureau. They were owned by—and used on a regular basis by—local communities for raising ducks, bathing water buffalos, washing clothes, and farming. While the alligators themselves are protected, and people can go to jail for killing them, there is nothing to prevent local farmers from draining the ponds to irrigate their crops or otherwise damaging the alligator's habitat. During the course of our surveys we came to realize that these areas were nothing more than the last few sites where, for a variety of almost accidental reasons, the last few groups of alligators had managed to hold on. Although the NCAR is large on paper, it is far from a wildlife reserve in any real sense of the word. It is a marker that designates the area where the Chinese alligator is making its final stand.

The marginal nature of the habitats in the NCAR places severe limitations on the survival of wild alligators. Alligators are squeezed between the dense human population in the lowlands and valley bottoms and the unsuitable biological conditions on the hillsides above. Ponds in the low valleys offer the kinds of habitats favored by alligators, but present intense human population pressures. While these areas have rich, organic soils and appear to support good populations of prey items (e.g., aquatic insects, eels, crayfish, snails, frogs), people live and work along the edges of the ponds on a daily basis. In some cases the ponds where alligators have survived are actually in the middle of farming villages. The hillsides above these ponds (especially in Xuancheng, Nanling, and Jinxiang counties) are unsuitable for agriculture and are used largely as pine or fir tree plantations.

The ponds themselves are tiny (the mean size of the sites we visited was 1.32 ha), and at one site the pond was only 0.09 ha (Changle). The cultivation of rice and other crops frequently extends up to the very edge of the water. In many cases there is no fringe of vegetation between the water and the cultivated fields, and at several sites (Zhongqiao, Zhucun, Yantang, Hongxin), shallow water sections of the ponds have been converted to rice paddies. At other sites (e.g. Shaungken, Yantang), the recent construction of water control structures has significantly altered the hydrology of the region, making them less favorable to aquatic wildlife. Under these conditions, conflicts between people and alligators are inevitable. Alligators must dig burrows to survive. Yet burrowing even a small distance into the shores of ponds will take alligators into agriculture fields, and moving between ponds requires walking through rice fields, another source of conflict with farmers.

The survival of alligators in these densely populated agricultural regions is a tribute to their biological resilience. They are small, relatively innocuous, and cryptic. They spend much of their time hidden in burrows and require only small patches of habitat to live. Their continued existence today in the NCAR also reflects the fact that alligators are long-lived, so even if successful breeding is no longer occurring, individuals may linger in an area for many years. A variety of site-specific habitat factors explain why alligators have survived in certain areas and not in others. The date of onset of agriculture and the intensity of agricultural development are certainly important. In some cases, including the Yanlin, Jiagu, and Wangjiameu designated sites, alligators are present only because the Anhui Forestry Bureau has used that area to release animals found in rice paddies or other inappropriate locations. At other sites, the survival of alligators is related to specific habitat factors that provided the alligators with some buffering from the intense human population pressures around them. At Zhangcun and to some extent Huangshugang, alligators may have been able to survive because a complex assemblage of former floodplain streams, now divided into numerous small ponds, provides a variety of hiding spots. Most critically, in areas where alligators are still found, their presence is usually tied to factors that have allowed them to successfully nest. At certain locations (Shaungken, Yantang) the stream or pond habitat used by the alligators is partially bordered by a forested, and relatively unused, hillside that provides nesting habitat. In other areas, small islands fulfill the same role (Hongxin, Changle, Huangshugang). Hill pond habitats are frequently surrounded by scrub vegetation or pine plantations that offer nesting habitat (Zhongqiao, Zhuangtou, Zhucun) with relatively little human disturbance.

In the preceding chapter we discuss the historical forces and trends that have shaped the nature of the lower Yangtze River basin landscape and its wetlands. In this chapter we describe the end result of that process as it pertains to the alligators by characterizing the nature of the last sites where alligators live in the NCAR. With the habitat description as an introduction, we then present the results of our surveys of the surviving alligator populations.

HABITAT CLASSIFICATION

Areas where remaining groups of Chinese alligators are found today are not representative of wetlands where they were distributed historically (Chapter 5). The sites where alligators remain are highly altered in a variety of ways by human activity,

Figure 7.2. Example of type 1 habitat, Heyi village in Nanling County.

either natural bodies of water that have been modified or artificial ponds created for water storage and agriculture. We classify these areas into three main types:

Type 1. Relict wetlands in low, broad, fertile valleys along the principal river courses. Areas dominated by agriculture, principally rice cultivation.

Type 2. Intermediate; ponds in tributary hill valleys but with significant agriculture in the valley above the pond.

Type 3. Artificial ponds created by damming streams in tributary hill valleys at the upper edge of rice or other crops or within the lower edge of tree plantations.

Type 1 habitats (Figure 7.2) represent the closest approximation to what was historically prime alligator habitat: low elevation, alluvial floodplains with a variety of wetlands including marshes, ponds, and streams. Today, these areas have been heavily settled and the associated wetlands have either disappeared or been significantly altered. The major rivers have levees along their banks and natural floodplains no longer exist (but the levees do break, leading to disastrous flooding). The former stream tributaries of the rivers that at one time wound across the lowlands have largely been diked and divided to create a series of ponds ringed with agricultural fields and villages. Natural shallow water bodies and marshes have been converted into rice paddies.

The few type 1 habitats with alligators represent tiny remnants of formerly extensive natural wetlands. They typically contain groups of small ponds (mean 2.9 ponds per site, average pond size 0.6 ha) in the midst of agricultural settlements. Despite

Figure 7.3. Example of type 3 habitat, the Zhongqiao reservoir, Jinxiang County.

their small size and the intense human use pressures that surround them, these areas support approximately half of the wild alligator population (Table 7.1). During our 2003 surveys, however, the percent of the observed and estimated population in type 1 sites fell to 35%. The sites we visited were either ponds that were formed by damming former streams (Huangshugang, Zhangcun, Heyi) or were the degraded remains of small natural ponds (Changle, Yantang, Shaungken). The soils in these areas were richly organic, and in many of the sites there was significant shallow water habitat and aquatic vegetation, both rooted and floating. At these sites we readily observed natural alligator prey (aquatic insects, crustaceans, snails, frogs, eels).

By contrast, type 3 sites (Figure 7.3) offer biologically marginal alligator habitat but reduced human land use pressures. These sites are characterized by oligotrophic (clear water) ponds with little aquatic vegetation or natural, shallow-water habitat. Like type 1 sites, these areas typically have multiple small ponds (mean 2.86 ponds per site, average size 0.95 ha) but are situated in hilly, partly forested terrain on a rocky soil that makes burrowing difficult. While alligators can survive in these ponds, they represent poor or marginal habitat, particularly for juveniles. At one type 3 site (Zhalin) that was used as a release site for alligators from other areas, we were told that many of the alligators abandoned the pond and were subsequently recaptured in rice paddies farther down the valley. In 1999, 26.1% of all alligators we observed were in type 3 sites, representing 25.3% of the estimated population (Table 7.1). These values increased to 30.0% and 40.5%, respectively, in 2003. However, as we believe there is a floating population of isolated individuals that were not observed during our surveys, and that most of these are likely hiding in small hill ponds, these values are likely to be underestimates.

Table 7.1 Survey results by habitat type (1999)

	No. of sites	% of sites	Area (ha)	% total area	Ponds	Ponds per site	Average pond size (ha)	No. of alligators seen	% seen	Estimated population	% estimated population	Estimated population density
							All areas visited					
Type 1	10	41.7	17.38	29.7	29	2.90	0.60	10	43.5	38	50.7	2.19
Type 2	7	29.2	22.04	37.7	8	1.14	2.76	7	30.4	18	24.0	0.82
Type 3	7	29.2	19.03	32.6	20	2.86	0.95	6	26.1	19	25.3	1.00
Total	24		58.45		57	2.38	1.03	23		75		1.28
							Designated sites					
Type 1	5	38.5	12.16	29.8	16	3.20	0.76	9	45.0	33	55.0	2.71
Type 2	4	30.8	19.20	47.0	5	1.25	3.84	7	35.0	16	26.7	0.83
Type 3	4	30.8	9.50	23.3	10	2.50	0.95	4	20.0	11	18.3	1.16
Total	13		40.86		31	2.38	1.32	20		60		1.47

Note: Table excludes two sites along rivers where it was impossible to calculate pond dimensions. Alligators were not present at either site.

Figure 7.4. Example of type 2 habitat at Yantan, Jinxiang County.

Type 2 habitats (Figure 7.4) are intermediate between the other two habitat types and are variable in nature, ranging from small ponds in rice fields to medium-size ponds in rolling hills surrounded by dry-ground agriculture. The two largest designated sites are type 2 habitats (Hongxin, Wangjiameu), and principally for this reason the mean pond size of type 2 habitats was more than twice that of type 3 sites (mean 1.14 ponds per site, average pond size 2.76 ha) (Table 7.1). Because of the size of these two ponds, plus the relatively large size of the Zhucun site, type 2 habitats represent nearly half of the surface area of protected ponds, but only about 25% of the estimated alligator population. Hongxin offers one of the best opportunities for undertaking habitat improvements with the goal of maintaining a wild group of alligators. The other large type 2 site (Wangjiameu) is one of the designated sites without alligators or adequate habitat. During the 1999 surveys, 30.4% of the observed alligators were in type 2 sites, representing 24.0% of the estimated total population. In 2003 the corresponding values were 35.0% and 24.3%.

Hillside Ponds: The Last Inadequate Refuge

The intensity of human population pressures in the agricultural valleys means that today, about 33% of the total area occupied by alligators, and 26% of the alligator population, is found in artificial ponds in low hills (type 3 habitat). While some of these ponds are relatively large, they receive direct runoff from hill streams carrying cool water. They have relatively clear water and little aquatic vegetation, either rooted or floating. While adult alligators can survive in these ponds, they may be less suitable for juvenile alligators, and the overall density of alligators is only half that of the type 1 sites (Table 7.1). Our impression was that these hill ponds are very unproductive biologically and qualitative observations of potential prey abundance suggest that food may be a limiting factor. At most sites the soil is rocky and has high clay content, making burrowing difficult. At one site (Zhongqiao), local farmers attributed the death of hatchling alligators to exposure to cold because there was no burrow near the nest site as a result of these soil conditions. While more investigation is needed, the combination of relatively cool water with little or no aquatic vegetation, limited food availability, and difficult conditions for burrowing may make these type 3 ponds harsh habitats for alligators. Under these conditions, the presence of alligators in these ponds reflects the fact that there is simply nowhere else they can go. Adult animals may be able to survive, and even nest, in these areas but there are real concerns about whether hatchlings and juvenile alligators can survive.

POPULATION STATUS IN THE
NATIONAL CHINESE ALLIGATOR RESERVE

Counting crocodilians is frequently a nocturnal activity. During the summer (when we conducted these surveys), alligators are mainly active at night and can be spotted by the telltale reflection of their eyeshine, even from a distance if a light is used. But if the animals are extremely wary, it may be easier to spot them during the day, floating on the surface of the water or resting on shore. Active burrows with fresh tracks are an unmistakable sign of the presence of alligators. By measuring the footprints left in the mud one can estimate the size of the animals. Local farmers are also a valuable source of information. These are people who have often lived next to the ponds for much of their lives, walking and working on a daily basis in areas where we were forced to make quick visits. Our description of the current status of the wild Chinese alligator population is based on the survey we conducted in 1999, as well as follow-up surveys in 2000, 2002, and 2003. Our initial survey in 1999 was the most thorough, and serves as a basis of comparison for subsequent surveys.

The numbers we present speak for themselves. The population of alligators outside of breeding centers is vanishingly small. More precise estimates of the number of alligators in the wild could have been obtained by a variety of methods, such as mark-recapture or standardized repetitive counts. But the few living wild alligators are already under a tremendous amount of stress, and intensive population surveys would have increased that stress, perhaps to the detriment of the surviving individuals. We began our surveys with the goal of coming up with an approximation of the total number of wild alligators, and that remained out objective. The extra time and money needed for more precise surveys were, we believed, better spent on other facets of the alligator's conservation. As a result, our population estimates may lack the rigor we would have wished for, but they do speak to the fact that the Chinese alligator is clearly a species on the edge.

Survey Methods

In 1999, field surveys were conducted from July through October at 26 localities within the five-county region of the National Chinese Alligator Reserve, including all 13 sites officially designated by the Chinese government for the protection of the Chinese alligator (here referred to as "designated sites") (Figure 7.1). We mapped sites using a laser rangefinder and compass, characterized the physical nature of the ponds and vegetation, and conducted interviews with local residents. Estimates of pond size in Table 7.1 are based on measurements made in 1999 when water levels were high and represent maximum values. Pond water levels were much lower in 2000 and 2003. At most of the 13 designated sites the Anhui Forestry Bureau employed a local farmer as a caretaker to oversee the protection of the Chinese alligators. The caretaker and other local farmers were usually good sources of information regarding the number of alligators, the locations of their burrows, recent nesting sites, and the history of habitat modification in the region.

We estimated the number of alligators at each site based on (1) nocturnal spotlight counts, (2) the presence of physical signs (active burrows, tracks), and (3) information provided by the local alligator caretaker and other farmers. Chinese alligators are extremely wary and difficult to observe, so we used the number of individuals seen as an estimate of minimum population size. The number of alligators estimated by the local caretakers and farmers were generally in agreement, and because they were based on almost daily observations made while working around the ponds were usually credible. We also questioned residents concerning alligator nesting and the location of active and old burrows.

We conducted nocturnal spotlight counts only if there was evidence that suggested Chinese alligators were present at that site within the last three years. Counts were initiated shortly after dark (ca. 20:00h) and typically were completed before 23:00h due to the small size of the areas surveyed. We located alligators by their reflected eyeshine using headlamps or a portable spotlight. Nocturnal surveys were conducted by either walking around the perimeter of ponds, using locally available wooden boats (two sites: Hongxin, Yantang), or rowing a two-person inflatable boat. When alligators were sighted, we approached as closely as possible to estimate total length (TL) in 50–cm size-class categories.

The sites we surveyed represent the largest known remaining groups of alligators. In other parts of the NCAR, farmers occasionally reported the presence of individual alligators to the local AFB office. Based on these reports, and a questionnaire survey completed by the AFB in 1998, the county forestry departments have estimated the number of alligators in areas we did not visit.

Using the results of the 1999 survey, we identified areas where alligators had been extirpated and did not include these sites in surveys in subsequent years. A total of 13 areas were revisited in 2000, 2002, and 2003. These sites represented the areas with the largest groups of alligators. They included all 10 designated sites where alligators remained and 3 other sites. Severe drought conditions limited the observability of alligators in 2000 when extremely low water levels resulted in most alligators remaining in their burrows during the survey period. In 2003 moderately low water levels as a result of another drought made direct observation of alligators difficult but facilitated the use of fresh tracks left by alligators moving in and out of burrows. We used the presence of fresh tracks to estimate the number of alligators present based on different size footprints.

Alligator population trends were determined by comparing the results of our survey with information on past nesting and previous estimates of alligator population size for individual counties and sites based on interview surveys and records provided by county offices of the Anhui Forestry Bureau.

Results of the Survey

During our 1999 survey we visited all 13 of the designated sites (Table 7.2). Although these areas represent the best sites within the remaining distribution of alligators, the total pond area was small (41 ha for 31 ponds in 1999). Furthermore, at three (23%) of the sites no alligators were present. The number of alligators at any one site was very small, ranging from 1 to 11 individuals. At the designated sites we observed a total of 20 alligators, and estimated the total number of animals to be 60 individuals. Overall population density was 1.5 alligators per ha. Alligator nests were found at 4 sites; in 1999, and only 2 (different) sites had evidence of successful reproduction during the last five years, as indicated by the presence of juveniles.

In addition to the designated sites, in 1999 we visited 13 other areas to survey for alligators. According to the AFB, these were the only other locations where alligators could be found. Excluding the 2 sites where pond area could not be measured (these were unbounded areas along the margin of rivers), there were 24 ponds totaling 17.6 ha. Only 6 of the 13 sites actually had alligators present. At these sites we observed 3 alligators and estimated the total number present to be 15, with group size ranging up to 5 individuals. The mean population density was 0.85 alligators/ha.

The size-class distribution of crocodilians provides information on the proportion of young and old individuals in the population. This information can be used to evaluate basic aspects of the population, such as a recovering population (a few adults with many juveniles) or one that is senescent (mostly adults with few juveniles). We were able to classify the size of 18 of the 23 alligators seen in 1999 (Figure 7.5). Only 5

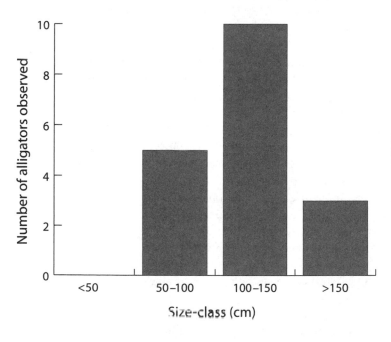

Figure 7.5. Size-class distribution of 18 alligators observed in the wild in 1999.

of 18 alligators (27.8%) were less than 100 cm TL, indicating a preponderance of adult alligators and a scarcity of juveniles. The lack of juveniles could be related to a small number of nesting females, high rates of egg mortality, high rates of hatchling alligator mortality, or any combination of these factors. The smallest animals observed in 1999 were 70– to 80–cm juveniles seen at 2 sites (Zhucun and Changle). In 2003 we found a group of 4 juveniles (60–80 cm TL) at Zhuangtou.

Overall, during the 1999 survey we observed a total of 23 alligators at 9 sites (Table 7.2). In 2003 we saw fewer alligators during nocturnal surveys (19) but were able to use lower water levels to confirm the presence of an additional 5 individuals based on fresh tracks, bringing the total number of confirmed alligators to 24. Based on these results and interviews with local farmers, we estimated a maximum number of adult and juvenile alligators present at each site (Table 7.2). Excluding the results of the 2000 survey, when water levels were extremely low, the number of observed alligators was 38% to 65% of the estimated total number. For 1999 and 2002, when the number of observed alligators was determined solely by nocturnal spotlight counts, the number of observed alligators was 41% of the estimated total. That number increased to 65% in 2003 when we also used fresh tracks to confirm the presence of alligators.

Combining all the results from our visits to 26 areas in the NCAR in 1999, we estimated the total number of alligators surviving at these sites to be 75 individuals. The total size of 55 ponds at these sites was 58.5 ha,, or an average population density of 1.28 alligators per ha.

Following our 1999 survey we published a report suggesting the total population of Chinese alligators in the wild was less than 130 individuals (Thorbjarnarson et al. 2001). This figure was derived from our estimates of the number of alligators that were found at the 26 sites we visited in 1999, plus an estimate of the number of single ("fugitive") individuals scattered across the landscape. The presence of this floating population of fugitive alligators is surmised from the irregular appearance of alligators in rice paddies and other areas around the NCAR, particularly in Xuancheng, Jinxiang, and Nanling counties. Little is known about these animals, but they may live as refugees in small waterholes or burrows in low hills and periodically move into agricultural areas to forage. When encountered, they are usually either captured and released at one of the designated sites or brought to the ARCCAR breeding center. AFB personnel report that in most cases animals released in designated sites do not remain there long and will move out again into surrounding agricultural areas. We calculated the number of these alligators based on information provided by the AFB, and the estimate represents, at most, an educated guess of the number of animals in this floating population. In 1999 we estimated there were a total of 55 animals living outside of the areas we visited (Jinxiang, 10; Xuancheng, 20; Guangde, 5; Nanling, 10; Langxi, 10). We currently have no reason to change these estimates; these alligators are all presumed to be adults (or nearly so) and (other than being killed by people) should experience low rates of mortality. But we emphasize that this figure represents an estimate of the maximum number of animals believed to be in the floating alligator population. Adding our calculation of the floating population (55) to the number at the known sites (75), our best estimate of the total number of wild Chinese alligators in the NCAR is 130 individuals, almost all of whom are adults. The

Table 7.2 Alligator survey results (1999–2003)

County	Map	Site	Pond area	No. of ponds	Protected	1999 Observed	1999 Estimated	2000 Observed	2000 Estimated	2002 Observed	2002 Estimated	2003 Observed	2003 Estimated	Juveniles	Last nest	Habitat type
Jingxian	E	Shaungken	1.71	6	yes	3	10	3	5	2	4	3	4	no	2007	1
	F	Zhongqiao	3.37	5	yes	1	4	1	4	2	3	2	6	yes	2007	3
	D	Yantang	2.53	1	yes	0	1	0	0	0	1	0	1	no	1992	1
		Liangzhong	1.07	3	no	0	0	n.d.	n.d.	n.d.	n.d.	n.d.	n.d.	no	1970s	1
		Linzhi	0.81	1	no	0	0	n.d.	n.d.	n.d.	n.d.	n.d.	n.d.	no	n.d.	1
		Ma'an	0.84	2	no	0	0	n.d.	n.d.	n.d.	n.d.	n.d.	n.d.	no	n.d.	1
Xuancheng	H	Yanglin	0.36	2	yes	0	1	0	0	0	1	1	1	no	1970s	2
		Maogia	1.25	2	no	0	3	n.d.	n.d.	1	1	1	1	no	1997	3
	G	Hongxin	8.44	1	yes	4	10	0	3	2	5	4	4	yes	2007	2
	ZT	Zhuangtou [b]	6.08	7	no	2	5	0	3	3	4	4	4	yes	2002	3
		Liyunmiao	0.67	1	no	0	1	n.d.	n.d.	0	0	0	0	no	1991	2
Guangde	M	Zhucun	2.76	1	yes	3	5	0	3	2	4	2	3	yes	1997	2
	L	Jiagu [a]	1.22	3	yes	0	2	0	2	0	2	1	1	no	1998	3
		Qianmouyuan	n.d.	1	no	0	0	n.d.	n.d.	n.d.	n.d.	n.d.	n.d.	no	n.d.	
Nanling	B	Changle	0.28	3	yes	5	11	4	4	3	4	3	4	no	2007	1
	A	Xifeng [a]	2.83	1	yes	0	0	0	0	n.d.	n.d.	n.d.	n.d.	no	1980s	3
	C	Zhalin	2.08	1	yes	3	5	0	3	2	4	2	3	no	1993	3
		Heyi	2.11	2	no	1	3	n.d.	n.d.	0	1	1	1	no	1970s	1
		Liudian	n.d.	1	no	0	0	n.d.	n.d.	n.d.	n.d.	n.d.	n.d.	no	1970s	1
		Shangma	2.2	1	no	0	0	n.d.	n.d.	n.d.	n.d.	n.d.	n.d.	no	n.d.	3
		Tianguan	1.67	1	no	0	1	n.d.	n.d.	0	1	0	1	no	n.d.	2
		Songcun	0.39	3	no	0	2	n.d.	n.d.	0	1	0	0	no	n.d.	1
Langxi	I	Wangjiameu [a]	7.64	1	yes	0	0	n.d.	n.d.	n.d.	n.d.	n.d.	n.d.	no	1950s	2
	K	Huangshugang [a]	5.35	4	yes	0	0	n.d.	n.d.	n.d.	n.d.	n.d.	n.d.	no	1983	1
	J	Zhangcun [a]	2.29	2	yes	1	11	n.d.	n.d.	1	5	0	3	no	1996	1
		Yuancun	0.50	1	no	0	0	n.d.	n.d.	n.d.	n.d.	n.d.	n.d.	no	n.d.	2
Designated sites			40.86	31		20	60	8	24	14	33	18	30			
All Sites			58.45	57		23	75	8	27	18	41	24	37			

Note: n.d. = no data available. Map symbols, which are only for designated sites, refer to Figure 7.13b. Pond areas were measured in 1999 during a high water period and represent maximum pond size. Pond size in 2000 was 20% that of the 1999 values. The date of last nest and the presence or absence of juveniles are measures of reproduction for each site.
[a] Removed from the list of designated areas in 2006.
[b] Designated an official area in 2006.

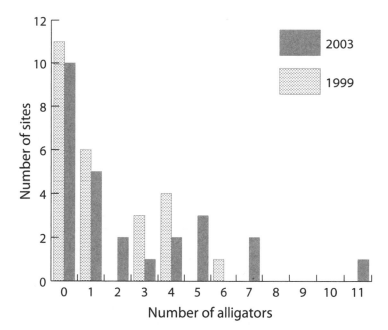

Figure 7.6. Frequency distribution of the estimated maximum number of alligators at each of the sites visited during the 1999 and 2003 surveys.

largest remaining groups have fewer than 10 individuals. In 1999 one site (Zhangcun) was estimated to have 11 alligators (Figure 7.6), but these were largely offspring from the last reported nest (1996), and during subsequent surveys we believe the number of alligators at this site has dropped to 3. The results of the 2003 survey suggest that the largest groups had declined to 6 individuals.

WILD POPULATIONS IN OTHER PROVINCES

Small numbers of Chinese alligators have been reported from sections of Zhejiang Province that are adjacent to the NCAR. Prior to our study no recent work had been done surveying wild populations in Zhejiang Province, but Z. Huang et al. (1985) reported that approximately 25 alligators remained in this area in the early 1980s. C. Huang (1983) had earlier estimated that the Lingfensi Tree Farm in Anji County had a population of 9 alligators (3 adults and 6 young). J. Fu (1994) indicated that alligators were spotted in 1991 and 1992 in Anji County, but the last record for Changxing County was a capture in 1979. After 1979 the only alligator reported from this region was one captured in 1998 in Anji County. This animal was placed in the Changxing alligator breeding center in 2000.

Today, the forestry department of Changxing County considers that alligators have been extirpated in the region (Z. Wan et al. 1998). No nesting has been reported in Zhejiang in recent memory. In August 2001 we carried out surveys at 7 sites in 2 counties (Table 7.3). While no alligators were observed, local farmers indicated that they had seen individual alligators at 5 different sites in Changxing County. This suggests that perhaps, a small floating population of alligators remains in parts of Zhejiang and that more survey work would be warranted.

Table. 7.3   Survey results from Zhejiang Province (2001)

County	Site	Water source	Alligators observed	Alligators reported by locals
Changxing	Maojiacun	river	0	1
Changxing	Xika	pond	0	1
Changxing	Er'jieling	pond	0	1
Changxing	Yanzimen	river	0	1
Changxing	Xikashang	2 ponds	0	1
Anji	Zhuzhongyuan	river	0	0
Anji	Xiaofeng	river	0	0

CHINESE ALLIGATOR POPULATION SIZE AND TRENDS SINCE THE 1980S

Although the NCAR was established in the 1980s ostensibly to protect the wild populations of alligators, it has failed to do so. Evidence of the continued decline of wild alligator populations since the establishment of the reserve comes from a variety of sources. Several attempts to quantify the total number of wild alligators have been made since the 1980s. But while all these surveys have agreed that the number of surviving alligators was small, they have been carried out using a variety of survey techniques and produced a range of figures that do not reveal any clear trend. Nevertheless, based on the estimates made in the mid- to late 1980s and our current estimates of population size, there is a clear indication that the number of wild alligators has declined. This is corroborated by information from individual counties and sites. Information about nesting, both in terms of the number of eggs collected by the ARCCAR from wild nests and the number of sites within the NCAR where nesting takes place, also provides unambiguous evidence of the loss of reproductive potential in the wild population.

Total Population Estimates

In the early 1980s surveys in Anhui Province estimated the wild population of alligators to be at least 300 individuals, comprised mostly of juveniles and subadults (Watanabe, in Groombridge 1982). Based on these and other surveys, it was calculated that approximately 500 alligators remained in the wild (B. Chen 1990a). Both these estimates, however, were based on incomplete surveys and likely underestimated the total number of animals. A more extensive survey of 129 villages (423 bodies of water) was organized by the AFB in 1985 and 1987 and estimated the number of wild alligators in the NCAR was approximately 735 (PRC 1992). As a total of 212 alligators had been previously collected from the wild to stock the ARCCAR breeding center from 1979 to 1983, it is likely that the total wild population in Anhui Province in the late 1970s was in the vicinity of 1,000 alligators.

In 1994, 77 alligators were observed in night spotlight counts during surveys conducted by staff from the ARCCAR, and the total alligator population in the surveyed areas was estimated to be 253, based largely on reports by local farmers (Table 7.4). Extrapolating to the entire 5–county region, the authors projected the total number

Table 7.4 Alligators in the Anhui Alligator National Reserve (1994)

	County					
	Xuancheng	Langxi	Guande	Jinxiang	Nanling	Total
Sites visited	7	6	5	7	6	31
Alligators observed	29	19	8	9	12	77
Burrows observed	18	6	11	36	33	104
Alligators reported by locals	68	47	29	56	53	253

Source: Survey results from C. Li et al. 1996.

of wild alligators remaining to be 667 to 740 (C. Li et al. 1996). But this extrapolation was based on the assumption that only 33% of the sites within the reserve were surveyed, even though they visited all the principal locations actually known to have alligators. The validity of this extrapolation is questionable and the total wild population estimate was most likely considerably overinflated.

A report by Y. Zhou (1997) presented total population estimates that ranged from 378 to 747 alligators in the NCAR from 1985 to 1994. The paper presented no information on the methods used and it is unclear how these population estimates were made. The 1990 figure (and presumably subsequent values) apparently includes 150 captive-reared alligators that were reportedly released, but no details are provided on this. Values from 1990 and 1992 also likely suffer from the bias in the 1994 survey. Z. Wan et al. (1998) indicated that in 1997 the population of wild alligators had been reduced to approximately 400, but no details of this population estimate were provided either.

There are many uncertainties in these data, but they do suggest that in the late 1980s and early 1990s the number of alligators in the NCAR was substantially higher than when we carried out our surveys. Overall, it is clear that from 1979 to 2003 the total population of wild Chinese alligators declined dramatically. Our best estimate of the magnitude of this decline is that over this 24–year period numbers fell from about 1,000 to approximately 100, a 90% decline.

County and Site-Specific Information

A variety of other information corroborates the dramatic recent decline in alligator numbers. Forestry Department records from Langxi County illustrate the reduction in the number of villages from which at least one alligator was reported (Figure 7.7). While the number of alligators at each of these villages is unknown, and in many cases reports probably represent single, displaced individuals, the shrinking area from which alligators are reported provides evidence of population decline. Today, alligators are known from only one area in Langxi County (Zhangcun). The average annual loss of villages in Langxi County reporting alligators from 1975 to 1999 was 4.0%.

In Nanling County, the local forestry department began conducting surveys of alligators in 1986; they visited a total of 66 villages in 16 towns. At that time just over 190 alligators were estimated to be present in 23 villages in 13 towns. That number has subsequently dropped precipitously (Figure 7.8). In 1998 the alligator population in Nanling County was estimated by the forestry department to be 32 animals in 9

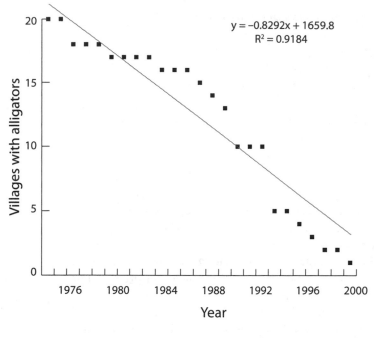

$$y = -0.8292x + 1659.8$$
$$R^2 = 0.9184$$

Figure 7.7. Number of villages in Langxi county where Chinese alligators were present from 1975 to 1999. Data from Langxi Forestry Department.

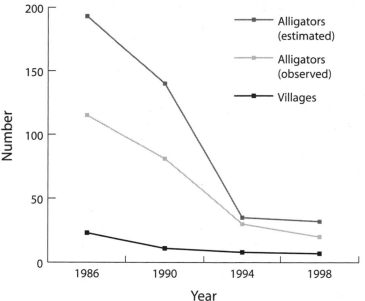

Figure 7.8. Alligator population trend in Nanling County since 1986. Data from Nanling Forestry Department.

villages. The average annual decline was 6.4% for the observed and estimated number of alligators and 5.4% for the number of villages with alligators.

Some information also exists for alligator population trends from direct comparisons of specific sites between 1981 (Z. Huang et al. 1985) and our 1999 surveys (Table 7.5). The overall decline of 84% during this 18–year period is equivalent to an annual population decrease of 4.7%. The overall trend for province, county-level, and site-specific population estimates and the number of villages reporting the presence of alligators indicates that the wild population is declining at a rate of 4% to 6% per year.

Reproductive Potential

As with the direct estimates of alligators, the yearly total of alligator nests, another good population index, has shown a significant drop over the past 20 years. Following the declaration of the NCAR in 1986, B. Chen (1990a) noted a sharp reduction in the number of eggs collected from the wild, indicating that nests

Table 7.5 Population decline of alligators at four sites

Site	1981 estimate	1999 estimate	% decline
Yantang	3	1	67
Shipu	4	0[a]	100
Yanlin	3	1	67
Heyi	22	3	86
Total	32	5	84

[a] Nanling Forestry Department staff reported that alligators have disappeared from the Shipu site.

were becoming scarce. During this period, the ARCCAR paid farmers to collect eggs (most came from Nanling and Xuancheng). The numbers fell appreciably from 1982 to 1989 (Figure 7.9) and remained low in the 1990s. The dates of last nesting at each of 17 sites visited during our 1999–2003 surveys also demonstrates declining reproductive output from the alligator population (Table 7.2). While in some cases these were imprecise descriptions (e.g., the 1970s), at most sites there was a good record of when the last nests were seen (Figure 7.10). Ominously, the decline has been particularly rapid since the mid-1990s, and in 1993, for the first time ever, no wild nests were reported (Figure. 7.11).

During our surveys, we found alligator nests in the wild at four sites. In 1999 we located three nests; one each at Shaungken, Zhongqiao, and Zhuangtou. Although we had intended to evaluate nesting ecology and the factors that determine nesting success, two of these nests were collected and taken to the ARCCAR breeding center. The nest found at Zhuangtou was left untouched, but failed to hatch, apparently as a result of low nest temperature (Chapter 5).

In 2000 one nest was found at Zhongqiao and left undisturbed. Of 19 eggs, 17 hatched and the hatchlings were subsequently observed near the nest. Eight of these were captured by the ARCCAR staff and taken to the breeding center but later returned to the site. No wild nests were found in 2001, but the presence of 4 juvenile alligators at the Zhuangtou site (68–80 cm TL) in August 2003 suggests there may have been an undiscovered nest there in 2001 or 2002. In 2002 and 2003 there were 2 confirmed nests, 1 each at Hongxin and Zhongqiao. By 2007 the number of wild nests had increased to 7, at 3 sites: 3 nests at Hongxin (2 of which were from females released there in 2003); 2 at Changle, and 1 each at Zhongqiao and Shaungken.

Capturing Wild Alligators

The ARCCAR breeding center was established in 1979 in the heart of the distribution of the remaining groups of wild Chinese alligators. Essentially, authorities believed there was little hope for the future of the wild population and set about collecting as many alligators as possible from the surrounding areas in order to establish a captive breeding center (Chapter 8). The AFB was generous enough to share with us the information concerning the origin of these animals and these data provide an opportunity to examine how the distribution of wild alligators has shifted from the valley bottoms to the hills over the last 25–plus years. A total of 212 alligators were collected from 1980 to 1988 (Figure 7.12), with the peak (79 animals) occurring in

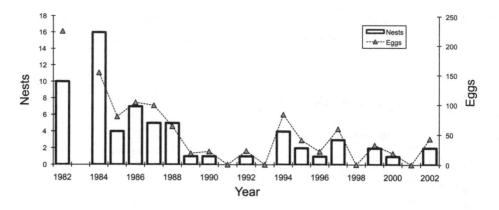

Figure 7.9. Number of nests and eggs collected from the wild in the NCAR, 1982–2002. The AFB stopped collecting eggs from the wild in 2003. Data from Z. Wan et al. 1998; AFB.

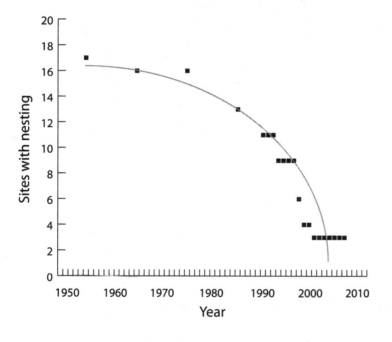

Figure 7.10. Decline in number of sites where at least one adult female Chinese alligator is present in the NCAR. Data for 17 sites visited during surveys; numbers for the 1950s to 1980s based on our surveys, interviews, and records from the AFB. Based on a starting point of 1955, the overall decline has been 82.4% over a 50-year interval (–1.9% per year).

1981. In subsequent years the number of animals captured dropped off sharply, suggesting that alligators were becoming scarce. Only 4 were collected in 1987 and 1 in 1988. Alligators were captured year-round, with a peak during the summer months (June–August); most were caught by farmers, who were paid based on the weight of the animals (Watanabe 1982).

The core of the NCAR is along the middle reaches of the Qingyi and Shuiyang rivers, and this is where all the ARCCAR animals originated (Figure 7.13; Table 7.6). The eastern section of the NCAR, in the Langchuan River drainage (Langxi and Guangde counties) still have a few alligators today, but none of the ARCCAR animals came from this area. We mapped the reported capture locations of 204 alligators (the capture location of 8 individuals was not well identified) and classified them into one of three habitats: valley bottom sites (= our type 1 habitats), intermediate

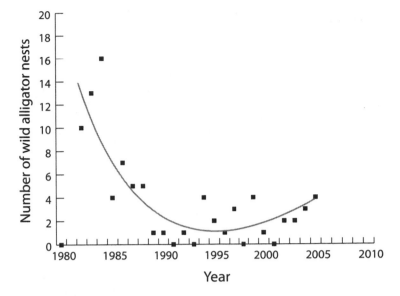

Figure 7.11. Total number of wild nests reported in the NCAR, 1982–2005.

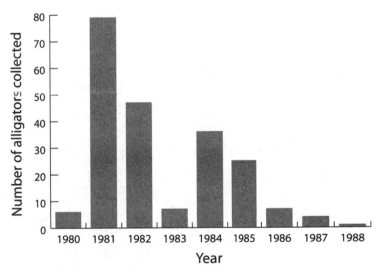

Figure 7.12. Number of wild alligators collected for the ARCCAR, by year. Data from the AFB.

(= type 2 habitat), and hillside ponds (= type 3 habitats). The majority of alligators (59.8%) came from valley bottom locations; 37.3% from intermediate sites; and only 6 (2.9%) from hill ponds (Table 7.6). We compared these values with the percentage of alligators observed in the three habitat types during our surveys (Figure 7.14). Distribution of animals among habitat types was not different between 1999 and 2003 ($\chi^2 > 0.9$, df = 2), but in both years results were very different from the locations of 1980s captures ($\chi^2 > 0.001$, df = 2). This difference was due to the dramatic increase of animals in the hillside ponds. These results may be biased if it were more likely that alligators from valley bottom sites would have been collected because they are the ones that typically are more conflictive for farmers. However, we believe this bias is minimal or nonexistent because the main incentive for collecting alligators was economic gain (Watanabe 1982) so farmers would also have captured known alligators

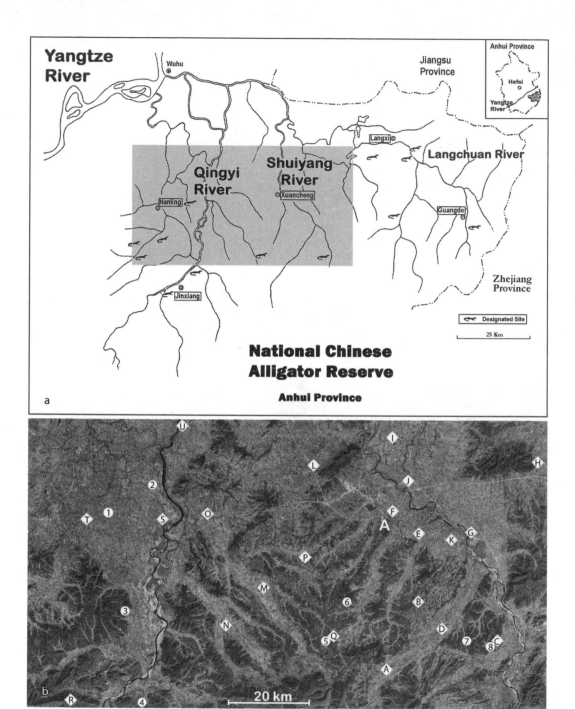

Figure 7.13. (a) The shaded area represents the area detailed in 7.13b; inset (*upper right*) shows the region with respect to the entire NCAR. (b) Locations of alligators captured in the 1980s to supply the ARCCAR. Central section of the NCAR shown as diamonds with letters (sites named in Table 7.6). Numbered circles are sites of principal locations where alligators survive today (1 = Changle; 2 = Heyi; 3 = Zhongqiao; 4 = Shaungken; 5 = Hongxin, 6 = Zhuangtou; 7 = Maogiatang, 8 = Yanlin. A is the ARCCAR. Heavily cultivated valley bottoms are shaded lighter than the forested hills.

Table 7.6   Alligators collected in the 1980s in the NCAR

County	Town or village (map location[a])	Number collected	Site classification	River system
Xuancheng	Xintian (A)	52	1	Shuiyang
	Yishan (B)	31	2	Shuiyang
	Yanglin (C)	27	2	Shuiyang
	Huangdu (D)	11	1	Shuiyang
	Banqiao (E)	2	1	Shuiyang
	Xiadu (F)	13	1	Shuiyang
	Sunbu (G)	1	1	Shuiyang
	Honglin (H)	8	2	Shuiyang
	Wuxing (I)	1	1	Shuiyang
	Shuangqiao (J)	1	1	Shuiyang
	Xiangyang (K)	7	1	Shuiyang
	Jingting (L)	5	1	Shuiyang
	Tuanshan[b]	4		Shuiyang
	Yangliu (M)	2	1	Qingyi
	Gaoqiao (N)	1	1	Qingyi
	Hanting (O)	5	1	Qingyi
	Jinba (P)	4	2	Qingyi
	Zhouwang [= Hongxin] (Q)	6	2	Qingyi
	Subtotal	181		
Nanling	Dafeng[b]	4		Qingyi
	Geling (R)	6	3	Qingyi
	Yijiang (S)	9	1	Qingyi
	Shipu (T)	3	1	Qingyi
	Subtotal	22		
Wuhu	Hongyang (U)	1	1	Qingyi
	Heping (NM[c])	6	1	Qingyi
	Wanzhi (NM[c])	2	1	Qingyi
	Subtotal	9		
Total		212		

Note: Site classifications: 1= valley bottom, 2 = intermediate valley hill ecotone, 3 = low hill ponds.
[a] Map symbols refer to Figure 7.13b.
[b] Location not identified.
[c] NM = not mapped; site is on the agricultural plain to the north of the NCAR.

from hillside ponds. This analysis suggests that over this 15– to 20–year period alligators continued to lose ground to agricultural expansion in the valley bottoms and were surviving in increasingly marginal habitats. The alligators were being "pushed" by the growing human population up into the hills and to their eventual extinction.

The increasing marginalization of alligators from the 1980s to the present can also be seen by comparing the collection sites with that of the current distribution of alligators (Figure 7.13). Many of the 1980s collection sites are in what would have formerly been the floodplain of the Shuiyang River. No alligators are found in this area today, and these groups may have been extirpated by capture for sale to the ARCCAR. Along the Qingyi River there are also floodplain sites that have disappeared, including the three most northerly (downstream) sites, in Wuhu County, two of which are to the north of the area shown in Figure 7.13.

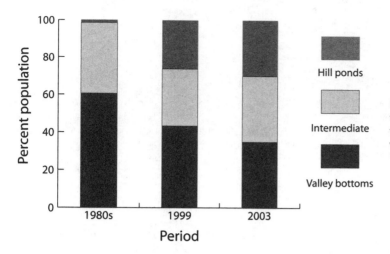

Figure 7.14. Alligators captured in three habitat types in the 1980s compared to those we observed during surveys in 1999 and 2003.

Moving south along these two river systems, one enters a series of hills that eventually ascend to the Huangshan Mountains. The farther south, generally the more marginal habitat as alligators move up these rivers into the hills and away from the alluvial plains of the Yangtze. Along this north-south gradient, latitude can be used as a proxy to measure the suitability of habitat. The average latitude of the 23 sites where alligators were captured in the 1980s was 30°54.07′ N. The average latitude of 11 sites where alligators were found in the Shuiyang and Qingyi river drainages in 1999 was 30°48.07′ N, a distance of 11.0 km farther south on average than the 1980s collection sites. This is yet another measure of how Chinese alligators have been losing ground—even since the creation of a national conservation program for the species.

Population Trend since 1999

Our survey in 1999 was the most complete and served as a baseline for the follow-up counts we did in 2000, 2002, and 2003. (Surveys were carried out in 2001 in Zhejiang Province. See below.) Because of drought conditions in 2000 we found the same number or fewer animals at all sites visited (Table 7.2). The total number of alligators directly observed in 2003 was 19; an additional 5 animals were counted based on tracks only. Surveys in 2005 were conducted by a team from the AFB and Anhui Normal University (Wu Xiaobing, pers. comm.; Table 7.7).

Changes in the number of alligators (observed or estimated) at each of the sites could have been caused by several factors:

   environmental conditions (e.g., water level)
   count variability
   successful reproduction and recruitment of juvenile alligators into the
       population
   immigration of alligators to the site from surrounding areas
   emigration of alligators from the site into surrounding areas
   deaths of alligators

The first two factors make interpretation of population trends more difficult as they do not represent actual changes in the number of alligators. Count variability is always a concern in wildlife surveys and may be a factor here, particularly as the 2005 surveys were carried out by a different team. Environmental conditions were a major factor, particularly in 2000 when a severe drought resulted in extremely low water conditions and many of the alligators remained inactive in their burrows for extended periods.

Alligators have continued to be killed at sites in the NCAR (Chapter 6), largely by accident, including both juveniles (captured in fish traps) and adults (accidentally poisoned or shot). In some areas where there are substantial conflicts between people (usually rearing ducks) and alligators, deliberate killing of alligators may occur on occasion. Immigration of alligators into the existing sites is unlikely given the large distances between the remaining groups. But emigration may be a factor as juvenile animals move out into the surrounding agricultural landscape and are either killed or join the floating population of alligators.

The lack of recruitment of juvenile alligators into the population has been a consistent feature of the recent alligator population decline. There are very few adult females in the wild population, and for these few individuals the odds are heavily stacked against producing surviving offspring. In some cases females remain as isolated animals without any nearby adult males. For those females that do have a mate, the availability of suitable nesting areas is extremely limited at most sites. The best nesting areas are usually on small islands, whose barrier of surrounding water limits human intrusion. But even when islands or other isolated sites are available, there may not be adequate vegetation for building a nest mound (e.g., the nest built of pine needles in Zhuangtou in 1999; Chapter 5). Egg predation in the traditional sense (natural egg predators) is not much of a problem. Over the last 20 years, however, eggs found by farmers have been collected and sold to the ARCCAR breeding center. This has resulted in the near total loss of hatchling production in the wild and is one of the principal reasons why so few juvenile alligators are seen. The few neonate alligators that do manage to hatch are faced with poor odds for survival beyond their first few months (which was a primary reason for egg collections by the ARCCAR). In recent years the two successful nesting sites appear to have had high rates of hatchling mortality. In Zhuangtou this may be the result of inadequate winter burrows in the rocky soil. In Hongxin water drawdown in the pond during late summer (to irrigate crops) results in extremely low water levels, which increases hatchlings' vulnerability to avian predators.

Alligators in valley bottom sites (type 1 habitat) continue to face intensive human use pressures and conflicts with farmers. One of the sites that had the best reported populations in the 1980s and 1990s, Zhangcun, has seen its number of alligators fall drastically. In 2006 it was taken off the list of alligator designated sites (H. Jiang et al. 2006), in large part as a result of conflicts with villagers who claimed alligators were eating their ducks.

The small population of wild Chinese alligators is highly fragmented and in a clear downward spiral. Our best estimate of the number of alligators at the main sites went from 75 in 1999 to 37 in 2003, a decrease of 50%. At present only three groups seem to have the potential for maintaining a stable population or, if habitat improvements are made, increasing in size (Figure 7.15).

Table 7.7 The 2005 NCAR alligator population survey

County	Township	Village	Location	Pond	Pond size (ha)	Individuals observed	Individuals reported[a]	Estimated alligator population
Nanling	Yidong (Hedong)	Heyi	Gangwanliu	Mengqian Tang	0.67	3	2	4–6
			Chengcun	Pu Tang	1.33	0	0	0
			Liujiaopan	Yangzi Gou	0.8	0	0	0
			Hejia	Louwu Tang	0.53	0	0	0
			Xiaotoucun	Xue Tang	0.4	0	0	0
	Jishan	Changle	Yangzhuang	Yangshu Tang	0.27	4	11–12	6–7
			Yangzhuang	Tulong Tang	0.07	0	0	0
			Yangzhuang	Duiying Tang	0.33	1	1	1–2
	Gongshan	Tian'guan	Gaocun	Shengyan Tang	0.67	1	1	1–2
	Daijiang Sanli	Kongcun	Wantangcun (Chalin)	Daiwantang reservoir	1.33	2	2	3–4
			Others					6
Jinxiang	Qinxi	Qinxi	Shuangken		3.13	2	2	3–4
	Changqiao	Zhongqiao	Tuanjie reservoir		1.3	0	0	0
					Marsh	0	5–6	6–7
					1.3	1	0	1
	Others					0	6	6
Xuanzhou District	Zhouwang	Hongyang (Bitian)	Hongshang	Hongxin reservoir	5.33	9	18–19	13–15
		Zhuangtou	Xishanfu	Jiujiaotang	3.0	4	2–4	6–7
				Sijiaotang	0.6	0	0	0
				Yaotang	0.23	0	0	0
		Jinba	Meicun	Eyutang	1.6	1	3	2–3

				Area			Reported[a]
Huangdu	Yanglincun	Yanglin forestry farm	Xiyitang	1.33	0	1	1–2
		Limucun	Maojiatang	4.5	0	1	1–2
	Xiangyang	Xiadu forestry farm	Fengchong reservoir	5.33	0	1	1–2
	Hanting	Qiucun	Dachong	0.33	0	1	1–2
	Honglin	Yangshu	Yangchong	4.0	0	1	1–2
		Daguocun	Wumutang	0.47	0	1	1–2
	Others				0	9	9
Langxi	Nanfeng	Zhangcun	Xiatang	2.33	1	2–3	2–3
	Others				0	4	4
Guangde	Lucun	Zhucun	Zhucun reservoir	1.18	0	0	0
		Zhongmingcun	Gongtang	0.13	0	0	0
			Hulutang	0.07	0	0	0
			Zhuchang reservoir	0.27	0	0	0
			Marsh	Marsh	4	8	6–7
			Marsh	Marsh	0	0	0
			Marsh	Marsh	0	0	0
	Shijie	Xiangshengcun	Lucun reservoir	230	0	2	2–3
		Jiagu	Dayan	1.13	0	2–3	2–3
	Others				0	4	4
Total					32	69–77	97–115

Source: Wu Xiaobing, Anhui Normal University.

Note: Surveys were systematized by breaking down locations into five organizational units from county to pond level.

[a] Reported alligators represent the number that local residents estimated to be present.

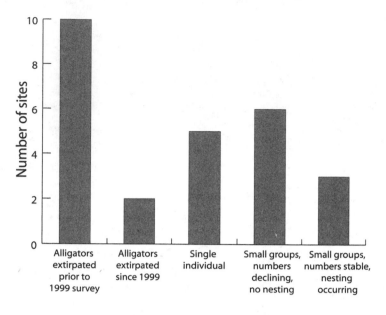

Figure 7.15. Summary of the status of Chinese alligators at the 26 principal sites in the NCAR. Maximum size of groups was 11 individuals.

In 2005 population monitoring was taken over by the Anhui Forestry Bureau in conjunction with Anhui Normal University. Surveys were carried out using the same methods that we used in 1999. Two new sites were visited (Tianguan and Meicun), both of which appear to have as many as 2 alligators. The overall population estimate was similar to 1999–2003 counts (Table 7.2), with the biggest increase at the Hongxin site, where three adult alligators were released in 2003 and where two nests hatched in 2004. In 2005 the upper estimate of total Chinese alligators alive in the wild was 120 (H. Jiang et al. 2006).

Changes in the Organization of the NCAR

In 2005 the Anhui Forestry Bureau contracted with the Academy of Forestry Inventory and Planning to undertake a review of the National Chinese Alligator Reserve and develop a master plan. The plan, which was adopted in February 2006, outlined changes in the size of the NCAR and the number of sites designated for the conservation of alligators. The overall size of the reserve was lowered from 433 km² to 308 km², and the number of designated sites changed from 13 to 12. The latter was accomplished by eliminating 3 sites where alligators had not been reported for many years (Huangshugang and Wangjiamen in Langxi County; Xifeng in Nanling County) or where only 1 individual had been seen (Jiagu, Guangde County). The last site stripped of its status as a designated site was Zhangcun, one of the sites that in the 1980s was reported to have the largest number of alligators, and is still believed to have 2 or 3 individuals. This site was dropped from the list because local farmers have complained vociferously about alligators taking their ducks. Two new designated sites were added, Zhuangtou, in Xuancheng County, which prior to 2004 was the only known site to have more than one nest in any one year. The other site is Gaojinmiao, Langxi County, where the Anhui Forestry Bureau has been releasing captive-reared alligators (H. Jiang et al. 2006).

While there was considerable debate about the pros and cons of these changes, in real terms they mean very little. The overall size of the reserve, initially 433 km², was always an arbitrary figure that merely represented the total area of the five counties over which the alligators were distributed. The new figure, although smaller, is equally arbitrary. The change in the designated sites likewise means little in terms of protection for the alligators. The lands at all the designated sites, with the exception of Gaojinmiao, are owned and managed by local farming communities and are not protected in any meaningful sense. Alligators, as a Class 1 protected species in China, still remain fully protected under the law. However, their habitat in the NCAR is not, and this remains the principal problem for developing recovery efforts for the Chinese alligator.

8

The Future of
the Alligator in China

The Chinese alligator faces significant conservation challenges as China undergoes unprecedented growth and change, metamorphosing from an agrarian society to one of the world's preeminent industrialized nations. Nearly one out every five people on the planet are Chinese and most of these live crowded into the low-lying, fertile plains in the eastern half of the country. While environmental degradation in this region is a problem that reaches back for centuries, the pace has quickened significantly since the 1980s. As in many developing countries, biodiversity conservation and other environmental issues take a backseat to policies that promote economic growth and the ensuing consumption of natural resources. China faces massive environmental problems that include desertification and loss of grasslands, deforestation, soil erosion leading to sedimentation and flooding, acid rain, and loss of wetlands (Mackinnon et al. 1996). Throughout Eastern China people are choking on the fumes from factories. More than 75% of the urban population live in areas that do not meet national air quality standards (M. Shao et al. 2006), and air pollution causes hundreds of thousands of deaths a year (Murray and Cook 2003). In China another basic element for human survival, fresh water, is scarce and frequently polluted. Worldwide, the per capita average availability of fresh water is 8513 m³ per year. In China it is less than one-quarter of this figure (Yang and Pang 2006). Water quality measurements in China have also found that the water supply in more than 25% of the country is not usable, even for agriculture (M. Shao et al. 2006). Air and water pollution issues will continue to grow in coming years as China's economy is predicted to quadruple in size between 2000 and 2020. Clearly, environmental contamination is one of the great issues that looms over the Chinese nation in the twenty-first century. Reining in the rampant growth that generates the environmental problems will require a delicate balancing act as it is precisely this spurt of growth that has allowed the Communist Party to deliver the gloss of prosperity needed to quell widespread civil unrest during this period of massive societal change. It is worrying to note that now, as environmental problems are reaching critical mass, most of the national pollution control benchmarks have gone unmet (Khan and Yardley 2007).

While the "brown" issues of air and water pollution are linked by everyday experience to human welfare, biodiversity and wildlife are much more intangible notions to people, be they peasant farmers or government officials. So if the political and economic changes that are required for improving the quality of such basic human requirements of air and water have not been forthcoming in China, what hope is there for species conservation? Wildlife conservation programs should aim to preserve ecologically functioning populations rather than the minimum necessary for the survival of the species. It is sad to note that in today's world there are some species for which this is no longer an option, and conservation efforts require a healthy dose of optimism, particularly in areas like Eastern China when so little of what is natural remains. Wildlife exists along a continuum that varies with respect to the degree of "wildness" of the environment and the level of human management necessary for survival. Among crocodilians, the spectacled caiman and American alligators, with large expanses of habitat and substantial wild populations are at one end of this continuum. The Chinese alligator, which survives in highly altered wetlands, is at the other. As habitat and animal populations dwindle, the amount of effort needed to ensure the survival of viable populations increases. In cases where the available habitat is very small some sort of long-term genetic and demographic management of populations will likely be needed. This is the situation with the Chinese alligator.

As a direct consequence of China's burgeoning population and long tradition of agriculture, habitat loss has had a devastating effect on the country's wildlife populations, an effect that has been significantly amplified by high levels of direct wildlife exploitation (Y. Li and Wilcove 2005). With few exceptions, most of the wildlife conservation news emerging from China has been sobering. Elephants, once widespread across much of southern China, survive as a few hundred individuals in southern Yunnan Province, with an increasingly female-biased sex ratio and a high proportion of tuskless males, the result of poaching (A. Smith and Xie 2008). Tigers are facing a similar fate; surveys over the last decade have failed to detect any individuals of the south China tiger (Tilson et al. 2004). In contrast, the giant panda, with considerable assistance from Chinese and international conservation groups, has managed to hold its own. While still one of the world's most endangered mammals, giant pandas have benefited from a well-financed multidisciplinary conservation program that has, in the last decades, placed considerable emphasis on habitat protection and has even achieved some habitat restoration (Lindburg and Baragona 2004).

Aquatic and wetlands species are similarly at risk throughout Eastern China. The *baiji*, or white flag dolphin, endemic to the Yangtze River has long been considered critically endangered. It was declared functionally extinct when an extensive survey failed to spot even one individual (Turvey et al. 2007). The loss of wetlands in the central and lower Yangtze River, one of the world's cradles of agriculture, has been particularly intense (Chapter 6). As a result of the dramatic decline in the size of wetland habitats, the numbers of wetland-linked species have been significantly diminished (S. Fang et al. 2006). And those species that remain are much more likely to be threatened. For instance some 50% of the birds listed as endangered in China inhabit wetlands (K. Chen and Zhang 1998). Conservation, and the restoration of wetlands, is sure to gain force in the coming years, if only as a means of protecting water supplies for human populations. This can already be seen in the wetlands parks around greater Shanghai. Some wetlands have been afforded a degree of protection in large part because they are used by migratory cranes, which are greatly esteemed in China. One example is the Poyang Lake in Jiangxi Province, where a network of protected areas has been established for wintering birds, particularly the Siberian crane (*Grus leucogeranus*). Another is the Yancheng Biosphere Reserve in coastal Jiangsu Province, home to a majority of the wintering population of the red-crowned crane (*G. japonensis*). The story of China's cranes and the giant panda give some cause for optimism: Two culturally significant species garnered enough political support in modern-day China for the implementation of effective habitat-based conservation efforts.

Like the elegant cranes, the Chinese alligator is a biological and cultural icon and a symbol of China's wetlands. It is China's sole surviving crocodilian and is linked, through a convoluted past, to its alter ego, the Chinese dragon, a quintessentially Chinese emblem that permeates society. As with the cranes, for the alligator to survive, wetlands habitats in Eastern China will have to be protected and restored. The Chinese alligator faces enormous, but not insurmountable, conservation challenges. Against all odds, alligators have managed to eke out an existence in the agricultural landscape that is the lower Yangtze River valley. We are optimistic that, given adequate support from the Chinese government, these numbers can grow and that viable wild alligator populations can be established in the relatively near future. Given adequate planning, the survival of Chinese alligators in select areas is entirely compatible with present and future land uses.

In this chapter, we discuss past and present conservation efforts for the Chinese alligator, and we describe our vision of what those efforts should be in the future. After summarizing previous work that has focused almost entirely on captive breeding, we address three main topics. The first deals with the issue of habitat, the most fundamental conservation impediment and the one that will be the most onerous to tackle. Ensuring adequate habitat for Chinese alligators will involve institutions and issues that go well beyond the rather narrow, but important, focus of alligator conservation. Second, we discuss the topic of reinforcements and reintroductions, that is, releasing captive-bred alligators to augment existing groups of wild alligators or to establish new ones. We believe that a network of small wild populations can be established and managed together with the captive groups as a "conservation metapopulation." Opportunities also exist where larger parcels of wetlands are being set aside or restored. These too can play a significant role in the restoration of the alligator. In our final section, we present possible approaches for the metapopulation management of critically endangered and habitat limited species.

BIODIVERSITY CONSERVATION IN CHINA

China is a huge country with a tremendous diversity of habitats, from coral reefs to the world's highest mountains, placing it among the world's most biologically diverse nations. Even by conservative estimates the country has more than 2,500 species of terrestrial vertebrates: 607 mammals, 1,244 birds, 384 reptiles, and 325 species of amphibians (Qian 2007), some 10% of the global total. Many, including the Chinese alligator, are found nowhere else. China is also the world's most populous country, with an estimated 1.3 billion people crowded mostly into the central and eastern parts of the country. Human population and land use pressures in these areas are enormous and there is little potential for establishing inviolate nature reserves. Habitat destruction has had predictable consequences for China's native fauna, contributing to the threatened status of 70% of China's endangered species (Y. Li and Wilcove 2005).

China also has a long tradition of environmental regulation, with some, such as the seasonal use of forests, rivers, and lakes, dating back to the twenty-sixth century BC (Watters and Wang 2002). But concrete actions to safeguard the country's wealth of biodiversity are relatively recent. The modern Chinese state dates to 1949, and its first provisional constitution (Common Program of the Consultative Conference of the Chinese People's Government) included references to conservation, but principally within a utilitarian framework: "Protect the forest and develop forestry in a planned way. Protect coastal fishing areas and develop fisheries. Protect and develop animal husbandry and prevent animal epidemic disease." At the third session of the National People's Congress in 1956, a group of renowned Chinese scientists provided the spark that initiated China's system of protected areas when they made a proposal to "designate specific areas in all provinces where the felling of trees is prohibited in the interest of conservation of natural plant life and scientific research" (G. Zhu 2002). That year, China's first protected area, the Dinghu Mountain Nature Reserve, was established. Over the next nine years the government of the People's Republic set aside areas that have since become some of China's best-known protected areas including Xishuangbanna State Nature Reserve in Yunnan Province, Wolong National Nature Reserve in Sichuan Province, and Changbaishan Biosphere Reserve in Jilin Province. During the socially turbulent period of the 1960s, little attention was given to the conservation of natural resources. This began to change in 1972 when China was admitted to the United Nations and participated in the UN Stockholm Conference on the Human Environment. Spurred by this event, China organized its own National Environmental Work Conference in Beijing in 1973 and created environmental protection agencies at the national and local levels. Between 1973 and 1978, a new set of protected areas was declared, bringing the total to 34, covering 1.3 million ha (Mackinnon et al. 1996).

In 1978 Deng Xiaoping took over the leadership of the Chinese state and initiated a period of significant change in Chinese society, including major economic reforms that paved the way for the development of modern-day China. Previously, wildlife policy in China had been principally concerned with exploitation (P. Li 2007), but in the post-Mao period it began a gradual shift toward conservation. The pace of environmental protection also quickened during this period, and in 1978 China worked with the UNESCO Man and the Biosphere Program to establish its first six biosphere reserves (Mackinnon et al. 1996). The following year, a trial environmental protection law established a general legal framework for regulating environmental degradation and promoted the rational use of natural resources. Also in 1979 the fifth National People's Congress passed an updated forestry law, which set specific goals for conservation of forest cover and included the basis for wildlife conservation programs. In 1981 the first ever PRC meeting to designate protected areas was held, and the following year a new national constitution was passed that stipulated, "The State ensures the rational use of natural resources and protects rare animals and plants" (W. Li and Zhao 1989).

Since the 1980s, China has set an ambitious agenda for creating new protected areas. Protected areas currently fall into different categories depending on the government level at which they are administered: national nature reserves, provincial

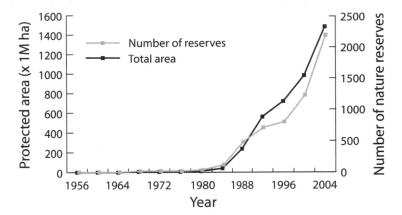

Figure 8.1. Increase in the number and size of protected areas and percent of China's total surface area since 1956. From State Environmental Protection Administration 2005; J. Xu and Melick 2007.

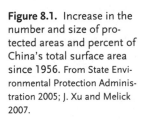

nature reserves, municipal nature reserves, and county nature reserves. By the beginning of the twenty-first century there were approximately 2,000 national and local nature reserves, covering 13% of the country's surface area, and plans call for increasing the number of reserves to 2,500 by 2050 (J. Liu et al. 2003) (Figure 8.1). But despite the rush to increase the amount of land under protection, there is no national plan for creating protected areas in China (G. Zhu 2002). And administration of reserves is shared by 10 different national level ministries or administrations. Consequently, responsibilities for the management of these areas are often unclear (J. Xu and Melick 2007). Overall coordination of reserves at the government level is assigned to the State Environmental Protection Administration, but most reserves are under the jurisdiction of the State Forestry Administration (in a different ministry). Other ministries responsible for protected areas are the Ministry of Agriculture, the Ministry of National Land Resources, and the Ministry of Water Conservancy. A recent evaluation of nature reserves carried out by the Chinese government found that one-third have no clear boundaries, staff, or management (J. Liu et al. 2003).

Over the last 20–plus years, the primary goal of the Chinese government has been to establish new reserves and increase the total area under protection, as opposed to increasing the effectiveness of existing reserves (J. Xu and Melick 2007). A good example of this is the NCAR, which appears substantial on paper but whose conservation value is, at best, minimal (Chapter 7). Many of China's nature reserves, including the NCAR, are in low-income areas, even by Chinese standards, and local communities and governments frequently see the declaration of protected areas as an impediment to economic development. Typically, the reserves themselves are inadequately staffed and underfunded. The total investment in national level reserve management in China ($113/km^2) is well below the worldwide average ($893/km^2) and even below the level of investment made by other developing nations ($157/km^2) (J. Liu et al. 2003). Additionally, much of the onus for generating funds for reserve management falls on the reserve staff itself, a situation that runs the risk of creating incentives for activities that are not compatible with the goals of the protected area (J. Xu and Melick 2007). In the case of the ARCCAR, economic pressures have forced the breeding center to become a local amusement park.

The responsibility for wildlife conservation in China resides largely in China's State Forestry Administration. Within the SFA, the Department of Wildlife Conservation has three main charges: (1) implementing conservation and management programs for China's wildlife, (2) managing nature reserves, and (3) coordinating programs for nationwide wetland conservation. China created its first list of "rare and precious species" in 1969, followed, in 1973, by regulations for the protection of wildlife. In 1983 the state council published a circular on the "stringent protection of endangered wildlife," noting that the protection of wild species was the duty of all citizens and for the first time banning the hunting or trade of endangered species (Watters and Wang 2002). The National People's Congress enacted its first law on the protection of wildlife in 1988, with the stated objective of "protecting and saving the species of wildlife which are rare or near extinction, protecting, developing and rationally utilizing wildlife resources and maintaining ecological balances." The law charges the forestry and fisheries administrations with the protection of terrestrial and aquatic wildlife, respectively, both at the national and the provincial levels. It also has provisions for the ownership of wild animals and addresses issues related to wild animal reserves; hunting; and the sale, purchase, transport, and import or export of state-protected wild animals and their products. This law also established a new system for classifying threatened species, with the most endangered listed as class 1, managed by the national government, and less threatened class 2 species, by the provincial governments.

A strategic plan for wildlife conservation in China, was developed at a meeting in December 2001. The result was a comprehensive plan for the protection of wildlife and the construction of nature reserves in the first half of the twenty-first century. The 2001 to 2050 program was divided into three phases, and an ambitious set of goals was delineated for the first 10 years (2001–2010), including specific benchmarks for wetlands protection and the creation of additional nature reserves (Box 8.1). One of the foremost goals was the recovery of a group of 15 species (and species groups) of fauna and flora. The Chinese alligator was listed as one of the "precious and rare" species.

Despite the bleak assessment of the status of its wild populations (Chapter 7), the Chinese alligator actually has many things going for it. As demonstrated so often, the Chinese people are capable of remarkable achievements. For the alligator, the biggest obstacle is not so much what can or cannot be done but the need for a paradigm shift in thinking about how to do it. With few exceptions, the entire focus of alligator conservation efforts in the past has been on captive breeding. Although there were some bumps in the road, this program has achieved notable success. The future lies in building on this success and, as part of wetlands conservation programs, restoring viable populations of alligators to areas outside of the breeding centers.

Crocodilians are an extremely resilient group of animals, something that has been well demonstrated not only by the survival of species like the Chinese alligator but also by the dramatic recovery of once-threatened species of crocodiles and alligators over the last three decades (Chapter 2). Captive breeding and reintroduction programs have played an important role in the recovery of some highly endangered species of wildlife, including crocodilians, and as habitats continue to dwindle will likely play an increasingly important role in the future. Over the last decade, the SFA has begun efforts to release captive-bred individuals of charismatic megafauna, most notably the giant panda and the South China tiger. The success or failure of these, and other, reintroduction programs boils down to one main issue—the availability of suitable habitat. Although the historic distribution of alligators is in the heart of the nation's rice basket, where land is at a premium, it has biological advantages over some of the other high-priority species in China. To begin with, it is relatively small (for a crocodilian) and cryptic, living much of its life in underground burrows, allowing individuals to survive almost undetected in agricultural landscapes. Their "cold-blooded" physiology, frequently denigrated as "primitive" in an evolutionary sense, is also a clear conservation advantage (Box 8.2). As ectotherms, alligators, and all other reptiles, have low energy demands when compared to mammals and birds of similar size. This translates into smaller land area requirements for individuals, and by extension, for viable populations. What this means is that a given piece of habitat can support larger populations of reptiles than it can of energy hungry mammals. For example, depending largely on prey densities, the home range of a single tigress can range from 15 km^2 in parts of India to nearly 400 km^2 in the Russian Far East. Given similar amounts of suitable habitat, entire populations of alligators can thrive in the space used by a single tiger. The advantages of this low-energy lifestyle are clear given the universal limitations on the amount of land that will be available for wildlife conservation in the future, and the demographic and genetic considerations that result from small population size. For conservation programs that rely on captive breeding it also pays to be a reptile: It is much more difficult to keep large numbers of mammals in captivity (you need much more space and food). Mammals are also more behaviorally complex than reptiles and typically require a greater investment of time and energy to maintain them in a psychologically healthy state that is important not only on ethical grounds but also to promote reproduction and the successful rearing of offspring. Mammals reared in captivity also usually require prerelease training in order to simulate natural learning of predator avoidance, foraging techniques, and so on (Kleiman 1989), behaviors that are more "hard-wired" in reptiles. The conservation advantages of being a reptile even extend to reproductive rates, which are typically much higher than for mammals of comparable size, based on a different life history strategy that stems from higher natural rates of juvenile mortality. In captivity, survival rates of offspring can be maintained at levels up to an order of magnitude higher than in the wild, resulting in the rapid growth of captive groups. China has invested considerable sums of money into programs to promote the captive breeding of the giant panda. In a good year 20 to 30 cubs are born, compared to 2,000 to 3,000 Chinese alligators a year in the main breeding centers.

The Conservation Advantages of Ectothermy

While from an evolutionary perspective crocodilians are more closely related to the endothermic "warm-blooded" birds, they share an ectothermic, low-energy approach to life with snakes, lizards, and turtles. As ectotherms, crocodilians maintain their body temperature at levels that allow normal activity by basking or by moving into areas of warm or cool water. Ectotherms generally have morphological or physiological adaptations to speed heat transfer between their bodies and the environment. But unlike endothermic mammals and birds they do not expend large amounts of metabolic energy from the food they eat to maintain an elevated body temperature.

Understanding the consequences of ectothermy is key to understanding the ecology and behavior of reptiles and how they differ from the more high-energy approach used by birds and mammals (Pough 1980). A low-energy approach to life is made possible by a low basal metabolic rate and a dependence on anaerobic respiration when sudden bursts of energy are needed. Because reptiles do not burn calories to heat their bodies, and because they tend to be less active than endotherms, they need considerably less energy. On a daily basis, a lizard uses 3% as much energy as a similar-size mammal (Bennett and Nagy 1977).

This low-energy ectothermic way of life means that reptiles can exploit ecological niches that are not available to mammals. Very small body size or elongate body forms result in high surface-area-to-volume ratios and high rates of energy (heat) loss to the environment. Ectotherms excel in living in environments where energy is limited (e.g., deserts) or where food availability undergoes extreme seasonal variability (such as during an annual dry period). Because they are less limited by food availability and individual energy requirements are considerably less, ectotherms can live at higher natural densities than endotherms. Because of their reduced metabolic rates, ectotherms are also much more efficient than endotherms at turning what they eat into body mass.

Ectothermy also has conservation advantages. Ectotherms do not need as much space to survive. Home range sizes of adult American alligators (typically < 30 ha for adult females) are an order of magnitude smaller than those of top mammalian predators such as tigers or lions. If we consider how much area is needed to support populations of a given size, the conservation advantages of ectothermy become even more apparent. Giant pandas have an average home range of approximately 8.5 km² or 850 ha. An area of this size of appropriate habitat for alligators could sustain some 340 adult female Chinese alligators (Box 8.3). Tigers are even more energy hungry than pandas. Reported home ranges for female Bengal tigers are over 40 km², an area that could potentially support a population of 1,600 adult female alligators (based on 2.5 ha per adult female). The small area requirements of ectotherms and the fact that they are much easier to raise in captivity (eat less, do not need as much space, are behaviorally more adaptable to captivity) make groups such as reptiles much more suited to conservation programs based on habitat restoration and reintroduction.

Conservation of the Earth Dragon

Wildlife conservation is about saving wild populations, not captive ones. Nevertheless, as habitats dwindle around the world, captive breeding can be an important tool in reaching this goal and the linking of wild and captive populations is a field that will gain increasing importance for wildlife conservation. In China, the creation of a captive breeding program has been the overriding objective of the official Chinese alligator conservation program, and relatively little attention has been given to the protection of the wild alligators. Our focus has been to promote the wild alligator agenda. But what exactly is a "wild" alligator and how is it distinguished from a captive one? While at first glance a seemingly inconsequential distinction, much of the direction of future conservation efforts for alligators in China will depend on making a clear distinction between wild animals and captive ones. The lack of clear differences between captive and wild groups is, in this case, indicative of a species in extremis. But we believe it also helps define some of the conservation challenges that the Chinese alligator faces and the issues that we as conservationists must address.

What Is a Wild Alligator?

Prior to our work on Chinese alligators, the need for criteria to distinguish wild from captive animals was a topic to which we had not given much thought. The difference between a "wild" and a "captive" individual seemed so straightforward there was little room for confusion. But during the course of our work we found instances where the boundaries between these two categories were blurred and we began asking ourselves what it was that differentiates a captive animal from a wild one.

At one end of the spectrum, we have the animals living in zoos around the world. For instance, the Bronx Zoo, operated by the Wildlife Conservation Society, has exhibited Chinese alligators for many years in New York City. These animals are housed inside a building in a relatively small exhibit, they are regularly fed, visitors pay admission to see them, and in the winter their environment is artificially heated so they will not die from the cold. These are clearly captive animals. At the other end of the spectrum are alligators living in a site like Zhongqiao. These alligators were born in nests by wild females living in ponds (albeit man-made ponds); they are not surrounded by fences nor are they fed. They fend for themselves, which includes digging their own burrows to survive the winter.

In China, there are also zoos that house Chinese alligators, as well as breeding centers that keep animals in more naturalistic surroundings. The ARCCAR, the breeding center in Xuancheng, Anhui Province, was established as a series of ponds in a small valley that reaches up into low hills covered with an old pine plantation, just the sort of habitat where many of the wild alligators, including those in Zhongqiao, are found. In fact, in terms of habitat quality, the heavily vegetated breeding ponds at the ARCCAR are superior to virtually all of the areas where we have observed wild alligators. A fence encircles these alligators and they are regularly fed by the ARCCAR staff. When the alligators nest, the eggs are collected and artificially incubated. During the period of cold winter temperatures, many of the animals are captured and brought into indoor, heated rooms; others remain in the ponds in burrows they have constructed. People pay to enter the center to see alligators. Consequently, although the ARCCAR has better habitat than that available at most wild alligator sites, it is clearly a captive population.

For many years the other main breeding center, at Changxing in Zhejiang Province, was known as the Yingjiabian Alligator Conservation Area and was declared a natural reserve. In fact, all the animals are within the walls of a breeding center where they inhabit man-made ponds and are regularly fed. The main difference between this breeding center and the ARCCAR is that at Changxing the eggs were left to hatch naturally in the nests. Once hatched, however, the juvenile alligators are removed from the breeding ponds and raised in separate pens, as they are at the ARCCAR. Here the issue was one of semantics. The breeding center was considered a "natural reserve" because the alligators had their own ponds, were protected from negative impacts of farming that went on outside the walls of the center, and were allowed to nest in a natural fashion. The young were separated so that their survival and growth could be maximized. In this sense they were far better off in the breeding center than if they had been left in the ponds outside. The essential underlying assumption was that there would be no present or future opportunities to improve or set aside habitat for the alligators.

The confusion between what is a wild alligator and a captive one is not, however, just a matter of semantics. It is a fundamental issue that goes to the root of the conservation philosophy for alligators in China. In 1992 the Chinese government prepared an official document for CITES, the international organization that regulates trade in threatened species, requesting the registration of the ARCCAR breeding center. As an officially CITES registered farm, the ARCCAR would be able to export alligators under the less restrictive requirements of CITES appendix 2. The document submitted to CITES included information on the status of the wild alligator population, showing that a recovery was well under way. Here the confusion between wild and captive animals comes into play (Chapter 1). At one site, Shaungken, the estimated number of live alligators was inflated in the report by adding animals that had been collected as eggs from nests at Shaungken but that were actually reared in captivity at the ARCCAR. In this crucial document the number of alligators living in the wild in the early 1990s at Shaungken was reported to be 82 and increasing at an annual rate of about 15%. When we visited this site in 1999 expecting to see more than 100 alligators, we were surprised when we could only find 3. It took us a while to determine where the difference in the numbers came from—virtually the entire "wild" population at Shaungken was in captivity at the ARCCAR.

Perhaps the most significant consequence of the failure to clearly differentiate wild from captive animals is the impression it has produced, at a national, policy-making level, that the Chinese alligator has been saved simply because the number of animals in captivity has soared. This is evident in many publications and press releases from the SFA. A June 2006 white paper from the State Council Information Office, titled "Environmental Protection in China (1996–2005), states, "The numbers of rare and endangered wild animal species, such as the Chinese alligator and red ibis, have increased by wide margins." However, the two species have very different conservation trajectories. The case of the red or crested ibis is actually deserving of praise. Once thought to be extinct in the wild, a remnant population of 7 birds was found in China in 1981, and as a result of conservation efforts the number of wild crested ibis in China is now in excess of 130 and increasing at a rapid rate. While the SFA has undertaken a captive breeding program for the ibis, it has also successfully fostered a remarkable recovery of the wild population. The same is not true of the Chinese alligator, where the increase in population size has been entirely within the captive population. The wild population has been largely ignored. Reinforcing this distinction in the minds of the SFA personnel and other policy makers in China is essential.

We started thinking more about what is a wild population of alligators and what is not one day at the ARCCAR breeding center. At the time, the facility held thousands of animals in a number of ponds and concrete grow-out pens. Occasionally, animals escaped from these pens and a few of the alligators were living in ponds outside the walls of the breeding center. These were now "wild" alligators, but as of yet there were no reports that they had been nesting. Inside the walls of the breeding center, the largest pond (called Alligator Lake) was no longer used as a breeding enclosure. Instead, visitors to the ARCCAR can rent boats to explore this pond or pay to catch fish. Approximately 10 escaped alligators have set up residence in this pond, and in 2005 there was even a nest on one of the islands. The question we raised was, Is this a wild population or a captive one? It was wild in the sense that the animals were not being fed, were nesting naturally, and were not subject to any of the other management practices that the captive alligators were (such as being brought indoors to overwinter). If this were to be considered a wild population, it would be the largest one known and would significantly increase the total number of alligators estimated to be living in the wild. But this group was captive in the sense that it was within the walls of the breeding center. We thought that perhaps another criterion for distinguishing wild from captive animals was whether people paid money to visit the pond where the alligators lived (we later decided this was of secondary importance). The best that we could do was conclude that this was a "semiwild" group of alligators and that it should not be added to the total estimated population of wild animals. We found it impossible to declare that a group of animals living within the walls of a breeding center, even if they are outside of the management activities of the breeding center, was a wild population.

The need for a shift in emphasis from captive breeding to restoration of wild populations is one we have been advocating since we completed our first survey of the wild population in 1999. In 2001 we helped organize an international workshop on the conservation of the Chinese alligator in Hefei, the capital of Anhui Province (Box. 8.3). The focus of this meeting was to advance plans to protect the minuscule wild alligator populations. There was considerable debate about the relative merits of potential sites in a number of different provinces, but the overall conclusion was that captive breeding of alligators in China had been successful but now a change in strategy was in order. The next phase of activities should emphasize protection of the current wild populations and establishment of new ones through habitat restoration and the release of captive-bred alligators. An action plan was drawn up with a specific set of goals.

Following the Hefei meeting, we were optimistic the Chinese government, particularly the SFA, would take the initiative and act on the recommendations. Unfortunately, the immediate outcome of the Hefei meeting was not a shift in focus to conservation of wild groups; it was, paradoxically, just the opposite (Box 8.6). Within months of the meeting there were large new cash infusions into the two main breeding centers to upgrade and build new pens at the ARCCAR and to build a new facility at Changxing. Restoration efforts for the wild populations were not undertaken because, it was claimed, there was no money available.

In 2008, more than seven years after Hefei, the Chinese government still had not taken the issue of habitat protection and restoration seriously. The one program that was to receive major support from the SFA and the Zhejiang Provincial government, under the guise of habitat restoration, was the creation of a new site that would become part of the Changxing alligator center. Initiated in 2007, this program began purchasing and landscaping 60 ha of land adjacent to the breeding center where Chinese alligators will be "released." But the site will be contiguous with the Changxing breeding center and encircled by a wall. Visitors will pay to enter the site to see the ponds and alligators, from a boardwalk or small boats. Some areas will be off limits to visitors, and we have been told that the alligators will not be fed. Nevertheless, we are left wondering how the alligators in this new site will be different from those in Alligator Lake at the ARCCAR. How can this be considered a wild population rather than merely an enlargement of the Changxing breeding center? One thing that will set it apart from Alligator Lake is the cost. The Chinese government will be investing an enormous sum of money (¥7 million or nearly $1 million), primarily to purchase the land and then landscape a new wetland. This is a huge investment for the creation of an adjunct to the breeding center, while the population of wild alligators continues to languish. The SFA continues to claim lack of funds for habitat restoration and reintroduction programs (H. Jiang et al. 2006), but its actions, as exemplified in the new Changxing project, indicate otherwise.

It is with a sense of irony that we note the history of alligators in the Changxing area, one of the last regions where wild animals remained in the 1970s. In the late 1970s and early 1980s local communities, fearful for the survival of the last alligators, and perhaps also with an eye toward using additional lands for agriculture, banded together to establish the breeding center. Alligators were captured from the local ponds and placed in the newly constructed pens within the walls of the facility, which was then declared an alligator reserve. After the international congress in 2001 to focus more attention on the needs of the disappearing wild population, the breeding center received a major influx of money from the government and was greatly expanded in size, becoming a local tourist attraction. Now, in a sense, the work is coming full circle, with alligators from the breeding center to be used for creating a new population. But the program trajectory has once again taken an unusual sidestep, by proposing the release of captive-bred animals into another expanded part of the breeding center, in the name of conserving the wild alligator population.

One might argue that the new Changxing "annex" is tied in with the development of a local ecotourism program and at least some of the funds come from sources tied to economic development, not habitat conservation. But this merely underscores the lack of support for conservation programs, even when they involve one of the 15 most "precious" species for the Chinese nation. There are other negative consequences of the Changxing project, particularly its effect on plans for the use of the nearby Xiazhu Hu wetland as an alligator restoration site. Xiazhu Hu offers an excellent alligator habitat, a large (by Eastern China standards) matrix of marshes and lakes in Zhejiang Province, and if used as an alligator release site it could support a large alligator population. However, one of the principal hurdles to obtain support from Zhejiang Province for the project has been the concern that if Xiazhu Hu has wild alligators it will diminish the Changxing breeding center's appeal as a tourist destination, which once again underscores the commercialism of the alligator conservation program. Will this continue to be the direction taken by authorities in Zhejiang? Will wildlife conservation in China be driven by its potential for economic return on investment?

History of Chinese Alligator Conservation

Chinese alligators were first recognized as a threatened species when they were added to the "grade A" list of protected animals in 1958 (State Forestry Administration Newsletter June 1998). This was, at best, a formality and resulted in no specific conservation actions

In 1972 a new list of protected species in China was promulgated and the alligator was listed as a class 1 protected animal, the highest protection level. Again, this resulted in little or no concrete conservation action, and alligator populations continued to plummet. Serious concern for the future of the alligator resulted in the first population surveys, carried out in the mid-1970s. In March 1976 Anhui Province set up the Wild Animal Resource Research Office, a group of experts led by two biology professors from Anhui Normal University, B. Chen and B. Li. The team carried out surveys in 11 counties in Anhui and estimated that there were no more than 400 alligators remaining in the wild. During the same period, C. Huang from the Chinese Academy of Sciences in Beijing was finishing surveys in Jiangxi, Jiangsu, Zhejiang, and Anhui provinces. His findings convinced him that the wild alligator population was in danger of extinction (State Forestry Administration Newsletter 1998). Outside of China, the lack of information about what was happening with Chinese alligators led some to speculate that the species might have gone extinct (Z. Huang and Watanabe 1986). In January 1978 an article in *National Geographic* magazine stated: "The Chinese alligator—no word in recent years on this only immediate kin of the American alligator; possibly extinct in the wild" (Gore 1978). This article brought the plight of the alligator to the attention of the Chinese government and prompted conservation programs for this beleaguered species (State Forestry Administration Newsletter 1998).

In August 1980 C. Huang visited the United States to attend the fifth working meeting of the IUCN Crocodile Specialist Group, in Gainesville, Florida. At that conference, he presented an overview of the status of the Chinese alligator and met Myrna Watanabe, who had just finished her PhD studying the behavior of American alligators. They began a collaboration that led to the first joint Sino-U.S. survey of Chinese alligators. With B. Chen, Watanabe and C. Huang visited seven areas from May to July 1981, conducting surveys that characterized the remaining habitat, estimated the number of wild alligators, and issued an urgent call for conservation actions (Z. Huang 1981; Watanabe 1982).

Captive Breeding of Chinese Alligators

Ancient records suggest that the Chinese may have been breeding alligators as early as 2600 BC. In the days of the legendary Huangdi (the Yellow Emperor) there were specialists who had great skill in breeding "dragons," a possible reference to "earth dragons," or alligators (Laidler and Laidler 1996). As wild populations of Chinese alligators and their habitat have dwindled, modern captive breeding programs have formed the basis of conservation efforts in both China and the United States. Building on the ancient tradition of breeding "dragons," modern-day captive breeding programs for endangered wildlife can serve a number of purposes: (1) establish secure breeding populations, (2) educate and engage the public in issues pertaining to the conservation of the species, (3) carry out research on the basic biology of the species, (4) serve as a focal point for fund raising-activities, and (5) provide animals for reintroductions (IUCN 1998). The Chinese alligator breeding program has focused principally on the first and third of these points. And due in large part to the alligator's adaptable nature, relative small size, and modest space requirements, captive breeding has been successful. Today there are a large number of alligators in Chinese breeding centers. In fact, most of the conservation focus in China has been devoted to two captive breeding programs, one in Anhui and one in Zhejiang province, the last two areas where wild alligators survived.

The first confirmed captive breeding of Chinese alligators actually took place in the United States in 1977, at the Rockefeller Refuge, Louisiana. It was part of a joint program between the Wildlife Conservation Society (then known as the New York Zoological Society), the National Zoo, and the Louisiana Wildlife and Fisheries Commission. WCS had tried unsuccessfully to breed A. *sinensis* at the Bronx Zoo from 1963 to 1975. In 1975 WCS entered into an agreement with the Louisiana Wildlife and Fisheries Commission to set up a breeding program at the Rockefeller Wildlife Refuge in southern Louisiana, in an area that approximated the conditions of their natural habitat in China. A pair of alligators that had been in captivity for 19 years at the Bronx Zoo were sent to Louisiana in 1976, along with a pair from the National Zoo in Washington, D.C. (captive ca. 37 years). Although one male died in 1976, the first eggs were produced the following year. The same female nested annually in subsequent years, and in 1980, the second female (from the National Zoo) also began nesting. The offspring from these animals were distributed to other zoos and by the mid-1990s the program involved 20 institutions in North America (Behler 1995). The North American captive breeding program for Chinese alligators became part of a species survival plan developed by the Association of Zoos and Aquariums.

In the 1970s, efforts were begun to rear Chinese alligators in captivity in China. In 1976 the Department of Biology at Anhui Normal University collected wild eggs and successfully hatched 16 alligators (B. Chen 1990a). The first successful captive reproduction in China was in the Shanghai zoo in 1979 (B. Chen 1990a). That same year both the Anhui and the Zhejiang captive breeding centers were established and these have since become the worldwide focal points for ex situ conservation efforts for the species. Today, these two breeding centers are extraordinarily important resources, containing the great majority of the world's population, and genetic diversity, of the Chinese alligator.

Anhui Breeding Program

On 2 January 1979 the Anhui Province government began work on a captive breeding center for Chinese alligators at the Xiadu Commune Tree Farm on land just outside the town of Xuancheng, the largest city in the heart of what remained of the alligator's range in China. The facility, which initially covered 89 ha of land in low hills, was supported entirely by the provincial government. In 1983 the breeding center began receiving support from the central government when Beijing provided ¥1.56 million (ca. $500,000) for construction and improvements. That year, the center expanded in size to about 100 ha and was renamed the Anhui Research Center for Chinese Alligator Reproduction (Watanabe 1982; H. Xu and Giles 1995).

The initial construction of the farm was completed in December 1979 and the breeding center began collecting a few wild alligators from the surrounding region. Little progress was made the following year, and in the spring of 1981, the facilities at the center were still very rudimentary and housed only 9 alligators (Watanabe 1981). The collection of wild alligators began in earnest during the summer of 1981. Local farmers were encouraged to capture alligators, most of which were excavated from their burrows, and were paid based on the weight of the animal. The farm also

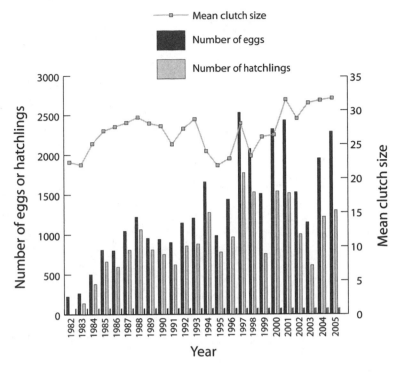

Figure 8.2. Production of eggs, hatchlings, and mean clutch size at the ARCCAR. Data from Z. Wan et al. 1998; Gu et al. 2000; and pers. comm. from the ARCCAR staff.

paid for each egg brought in from wild nests, even though the vast majority of eggs were dead by the time they reached the farm (Watanabe 1982). From around the region farmers arrived at the gates of the breeding center with hog-tied alligators and baskets full of alligator eggs (Z. Huang and Watanabe 1986). By the end of July 1981, there were 89 animals on the farm, and by the summer of 1982 this had risen to 130 to 140 alligators, mostly adults (Watanabe 1981, 1982). While most of the alligators came from the five-county area that today comprises the NCAR, some came from farther afield (Chapter 7). In June–August 1983, after extensive flooding of the Jiang-nan plains near Wuhu, farmers captured more than 50 wild alligators and brought them to the farm (State Forestry Administration Newsletter 1998). These must have been among the last of the alligators from this historical stronghold of the species. The total number of wild captured alligators brought to the farm was 212 (PRC 1992; Webb and Vernon 1992), of which the vast majority (95.4%) were adults (B. Chen 1990a). Approximately 1,400 eggs were brought in by farmers or collected by the ARCCAR staff from wild nests between 1982 and 2002.

Initially, the ARCCAR had significant problems keeping the alligators alive and getting them to breed. Watanabe (1981) describes early efforts to house the alligators and the lack of breeding success, partially a result of limited access to information on crocodilian breeding programs in other countries. However, between 1982 and 1984 the ARCCAR worked out appropriate husbandry protocols and had significant success breeding alligators (Figure 8.2). Records of the number of eggs incubated and hatched during the first few years of the farm operation are difficult to interpret as most accounts do not differentiate between eggs collected from wild nests and eggs produced on the farm by captive females. In 1981 the first alligators hatched at

Figure 8.3 (*facing page, above, and left*). The Anhui Research Center for Chinese Alligator Reproduction, in Xuancheng, southern Anhui Province: (a) alligators being returned to breeding ponds after overwintering indoors; (b) a breeding pond set in a natural valley at ARCCAR; (c) a group of subadult alligators basking; (d) one of the authors (XW) viewing eggs in an incubation chamber.

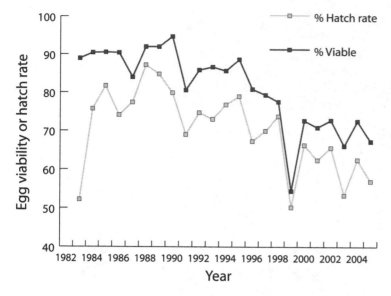

Figure 8.4. Rates of egg viability (percentage of eggs that develop embryos) and overall hatch rates (percentage of eggs laid that produce hatchlings) at the ARCCAR from 1983 to 2005. Data from Wan et al. 1998; Gu et al. 2000; and pers. comm. from the ARCCAR staff.

the ARCCAR when 40 young emerged from more than 200 eggs. All of the hatchling died the following year (Z. Huang and Watanabe 1986) and it is likely these were eggs collected from wild nests and not produced on the farm. In 1982 there were reportedly 270 eggs collected from wild nests, of which 224 were incubated and 147 hatched (Webb and Vernon 1992). It seems likely that the first nests produced in captivity were in 1983; in that year 278 eggs from wild nests were also collected (B. Chen 1990a). In 1984 the first of the more natural habitat breeding enclosures at the ARCCAR was completed (Crocodile Specialist Group 1984) and captive breeding increased dramatically, with the first F_2 offspring reported in 1988. During the late 1980s and early 1990s the center's production stagnated (Figure 8.2). This was during the time that Webb and Vernon (1992) visited the ARCCAR and reported that budget problems were a major constraint. From 1994 to 2000 the number of eggs and hatchlings increased significantly, leading to problems associated with an oversupply of alligators. For a short period after 2000, the production of nests and eggs dropped substantially (Figure 8.2). At the same time the rates of egg viability also fell (Figure 8.4). This decline, for which no reasons were identified by the ARCCAR, was worrisome given that the ARCCAR is by far the largest repository of captive Chinese alligators in the world. Production reached a low in 2003. Since then nesting has increased and is approaching levels seen in the late 1990s (Figure 8.2).

The main breeding areas for alligators at the ARCCAR are a series of ponds formed by damming a small stream that runs through the facility. After the alligators nest, the eggs are collected and incubated at about 30°C to produce a female-biased sex ratio. Young are separated by year class and fed mostly on fish purchased from a local market. Hatchlings are kept in a group of concrete grow-out pens with some aquatic vegetation. As the captive population at the ARCCAR has grown, an ever-increasing number of pens for larger juveniles and subadults have been built.

The ARCCAR has a director, an assistant director, five research staff, and approximately 35 employees who tend to the alligators. The ARCCAR was registered as a captive breeding center for Chinese alligators with CITES in 1992. The registration is significant because the Chinese alligator is listed in CITES' appendix 1, which prohibits commercial trade; but because the ARCCAR is registered with CITES, alligators produced there can be traded internationally under the less restrictive regulations of appendix 2. The international trade of alligators and alligator products was initially seen as a way to generate funds to help support the center's operation. However, trade has not been a significant source of income due to a lack of demand for live Chinese alligators or for their skins or meat. The ARCCAR does generate funds through local tourism. In 1996 the facility was transformed into a small zoo and recreation center with a children's playground (Z. Zhang 1996), fishing, and even an alligator "wrestling" show. Entrance fees and income from concessions in the park support the ARCCAR. In 1997 we were informed that annual operation expenses were approximately ¥1.2 million (ca. $150,000). At that time, 70,000 to 80,000 tourists visited the park annually (70% from Xuancheng County), generating approximately ¥700,000 to ¥800,000 ($85,000–$97,000) in revenue (Z. Zhang 1996), which covered 40% of the center's expenses. The sale of alligator meat domestically has been very limited, largely out of concerns that China would be seen as commercializing a critically endangered species (X. Meng, pers. comm.). A small number of restaurants in China have been approved for the sale of Chinese alligator meat, including ones in Xuancheng (Figure 6.8), Hefei, Huangshan, and Beijing. In 1999 the Xuangcheng restaurant was selling alligator meat for ¥2,000 per kg ($250/kg), but monthly sales were only 7.5 kg. In 2007 the center appeared to have adequate funding, judging by the amount of construction. This was especially notable following a major infusion of support from the SFA in Beijing as a result of the 2001 conference in Hefei (Box 8.6).

Changxing Alligator Breeding Center, Zhejiang Province

The Changxing Alligator Breeding Center (CABC) was established in 1979 (the same year as the ARCCAR), in Changxing County, Zhejiang Province, 92 km E of the ARCCAR. Initially, the CABC operated with funds from local governments, along with donations from nearby universities, local groups, and individuals. Originally called the Yingjiabian Alligator Conservation Area, it consisted of a 0.67-ha walled-in facility containing 10 breeding and rearing ponds totaling 2,400 m^2 (J. Wang and Huang 1997). Older residents report that in the 1950s more than 100 alligators were captured annually from the region (J. Wang and Huang 1997). The area was the principal source of alligators for the Shanghai Zoo, some of which were subsequently traded internationally. Although C. Huang (from the Chinese Academy of Sciences in Beijing) was a scientific adviser in the program's early years, it was a locally based operation until November 1988 when it was officially recognized by the Zhejiang Province Forestry Department and began receiving support from the provincial government. The facility obtained a breeding permit from the National Forestry Ministry in September 1992. The goals of the project have been to establish a successful breeding program, create an alligator lake to promote tourism, and conduct research on the utilization potential of alligators (J. Wang and Huang 1997).

The original breeding stock came from nearby areas in Zhejiang Province. M. Huang et al. (1987) noted that at least three came from Anji County where 9

alligators (3 adults and 6 young) were reported to be living on the Lingfengsi tree farm in 1983. The number of founding breeders is not clear; J. Wang and Huang (1997) reported that the farm initially had three animals; Webb and Vernon (1992) reported four animals were collected; but we were told that by 1982 the center had 12 animals, the 3 from Anji County and 9 others (4 males and 5 females) that were captured in the area around the site in Changxing County (Crocodile Specialist Group

Figure 8.5 (*facing page and above*). The Changxing breeding center, Zhejiang Province: (a) the main alligator exhibition pond; (b) an adult female alligator basking on shore; (c) entrance to the breeding center's exhibition pond.

1983). As with the ARCCAR, husbandry problems resulted in high mortality rates during the first years of operation. J. Fu (1994) reported that 8 adults died in 1982. The first clutch of 20 eggs was produced in July 1983, but all the eggs were infertile. The following year the first fertile clutch produced hatchlings that emerged on 13 September 1984. F_2 alligators were produced for the first time in 1997 (S. Fang 2001). Throughout the late 1980s and early 1990s, annual reproduction was limited to 1 to 2 nests (Webb and Vernon 1992). In 2006 there were 13 nests. Unlike in the Anhui center, eggs are left in the nests until they hatch, then the hatchlings are collected and placed in grow-out pens. As of 2007 the total number of alligators at the Changxing center was about 300, including 100 adults purchased in 2006 from the ARCCAR.

Other Captive Alligator Groups in China

In addition to the two main breeding centers, a number of zoos and other facilities in China maintain relatively small populations of Chinese alligators (J. Fu 1994). The first captive breeding in China took place in 1979 at the Shanghai Zoo. Chen (Crocodile Specialist Group 1983) reported that 5 adult females had been nesting in the late 1970s and early 1980s. This was followed by successful breeding in 1983 at the Ningpo Zoo in Zhejiang Province. Other zoos that bred Chinese alligators in the 1980s were the Beijing Zoo and the Chengdu Zoo. By the 1990s offspring from the captive breeding programs began to spread to other facilities in China. The CABC was reported to have sold 10 alligators in 1991 to Qiandaohu National Park Forest in Zhejiang Province, 7 alligators to the Hangzhou Aquarium, and 3 to Yuhang Park (both in 1992). Alligators from the ARCCAR have been used to establish a number of other facilities. In 1990 an alligator farm was set up in Baiteng Lake, Zhuhai, Guangdong Province; in 1992 it was reported to have 70 A. sinensis (Z. Zhang 1992). In 1991 an alligator breeding facility was also opened in Da-zhi-po, Qiongshan, Hainan Province with 50 Chinese alligators (Z. Zhang 1992). In 1994 this facility produced 20 eggs but had a low viability rate. Similar problems were reported in 1995 (Z. Zhang 1995). In the mid-1990s the Beijing Crocodile Lake Park was reported to have about 300 A. sinensis (J. Wang and Huang 1996). The ARCCAR also sold 200 alligators to the Guangzhou English Special Animal Experimental Farm in 1992 (J. Fu 1994). By 1995 the Guangdong population was producing eggs (95) but was experiencing low fertility rates (only 2 hatched; Z. Zhang 1996). As with the group on Hainan this was largely attributed to the climate in southern China (Z. Zhang 1996). In 2006 and 2007 the ARCCAR began selling a large number of alligators to a private individual in Nanjing who was establishing what may become the largest breeding center for Chinese alligators. Reportedly, the ARCCAR has a contract to sell 10,000 alligators to this new center over the next decade (C. Wang, pers. comm.).

THE QUESTION OF HABITAT

Unquestionably, habitat loss is the principal reason why the Chinese alligator is on the brink of extinction in the wild. Without wetlands, there can be no alligators, and while wetlands have long had a bad reputation as breeding grounds for disease and unsavory animals, they are essential components of the biosphere for humans and other animals. Historically, the middle and lower Yangtze landscape harbored one of China's greatest concentrations of wetlands, including a vast array of freshwater lakes and marshes associated with the river's floodplain. Today, the Yangtze is not so much a living river as it is an engineered canal, and agricultural and urban settlements have mushroomed on former floodplains as a result of intensive land modification. Floodplain lakes and marshes have disappeared or shriveled after the construction of thousands of kilometers of levees and dikes cut off the river's links to lakes (Chapter 6). These areas used to form a complex wetland network that performed important natural functions such as spawning and feeding for fish and other aquatic fauna.

Although China has a long history of wetland loss (Chapter 6), in recent years there has been a growing awareness, in China and across the globe, of the environmental costs associated with disappearing wetlands. The loss of wetland floodplains, which naturally act as sponges to absorb and store excess water, was one of the contributing causes of the disastrous floods along the Yangtze River in 1998, floods that resulted in significant national policy changes concerning the protection and restoration of wetlands. Water quality issues are also moving to the forefront of debates concerning development in China, and this is one of the major forces behind the creation of a series of wetlands parks in the Huangpu River drainage in Shanghai municipality. The future trajectory of wetlands conservation programs in China will determine much of the potential for restoration of populations of aquatic wildlife and is an important topic for consideration in the development of long-term conservation plans for Chinese alligators.

Wetlands Protection

Exacerbating the weak system for protected areas in China (Chapter 8) are a series of problems linked specifically to wetlands conservation. Despite the tremendous amount of land converted for agricultural purposes, China still has a significant amount of wetland habitats, including an estimated 12 million ha of lakes and 11 million ha of marshes (compared to over 38 million ha of rice paddies) (Scott 1989; J. Chen and Chen 1997). But the conservation of wetlands in China has suffered from a lack of public awareness of the importance of these habitats and the array of problems they face. This is aggravated by weak protective legislation, a lack of funding and management expertise, and a booming economy that is filling and polluting wetlands.

J. Chen and Chen (1997) characterized the greatest problems for wetland conservation in China as a lack of public understanding of the functions and values of wetlands, inadequate legislation and funding for protecting wetlands, and wetland destruction and degradation resulting from population and economic growth. The year after this analysis was published, nature saw fit to highlight some of the consequences of wetlands destruction. In the summer of 1998 the Yangtze River overflowed its banks with catastrophic results. Heavy rains combined with the landscape's decreased capacity to retain floodwater because of deforestation and the loss of wetlands resulted in the river breaching its levees in numerous locations along the middle Yangtze. An estimated 223 million people were affected by the flood and nearly 5 million houses were destroyed (Zong and Chen 2000). Thanks to early warning systems and the government's coordinated response, the death toll was less than 4,000 (compared to 145,000 in a similar event in 1931), but the flood resulted in widespread economic devastation.

Following the 1998 Yangtze flood, the Chinese government assumed a more proactive role in addressing the loss of wetlands, making the protection and wise management of wetlands a priority. A circular issued by the State Council General office (5 June 2004) requested that all levels of government balance development with science-based management of wetlands and that the SFA develop a national wetland conservation plan. Some of the most important changes put into place following these floods were a logging ban in the upper catchments of the Yangtze, removing steep cultivated slopes from agricultural production, increasing reforestation efforts, relocating villages from flood-prone areas to higher ground, and restoring wetlands in the Yangtze floodplain. These are huge undertakings that count on vast resources from the Chinese government. The conversion of hillsides now under cultivation to forest (Sloping Land Conservation Program) has as a goal the conversion of nearly 15 million ha of agricultural fields (on slopes greater than 25 degrees) to forest by 2010 and has a project budget of ¥337 billion (more than $40 billion) (J. Xu et al. 2004).

The Chinese government is also counting on the Three Gorges Dam and other massive infrastructure projects to play a significant role in water retention and flood alleviation in the future. From 1998 to 2002 the government spent ¥165 billion ($20 billion) to repair and strengthen levees and other flood control measures in the middle Yangtze. For the first time this also included the deliberate reflooding of agricultural lands to act as buffers against flooding. The government spent ¥10 billion (ca. $1.2 billion) to destroy polders and relocate the people living in the newly flooded areas. This is primarily taking place in the area of Chenlingji (flooding equivalent to 21 billion m³), Wuhan (6.8 billion m³), and Hukou (4.2 billion m³) (H. Wan 2003). In 29 counties in Hunan Province around Dongting Lake a total of 333 polders were removed between 1998 and 2002, and 558,000 people were resettled.

Other large-scale wetlands conservation-restoration projects are currently under way in China. The PRC-UNDP Global Environmental Facility project "wetland biodiversity conservation and sustainable use in China" focuses on four wetlands complexes, two of which are in the historic range of the Chinese alligator (Yancheng coastal marshes and Dongting Lake). Another large-scale initiative is a Dutch-funded program administered by WWF-China, the Living Rivers program on wetlands and biodiversity conservation, which focuses primarily on the Dongting region. A related program by WWF-China in association with HSBC the Hong Kong Shanghai Banking Corporation aims to find natural management solutions for the Yangtze that will benefit people and wildlife.

Options for Habitat Restoration

If the future of Chinese alligators outside of captive breeding centers depends on reintroduction, the first question that arises is where can alligators be released? The designated sites in the NCAR are all, at best, marginal and will only support small breeding groups for the foreseeable future. While improvements of some of these sites and the incorporation of these alligators into a larger conservation metapopulation should remain a high priority, other options outside of the current limited range of the alligator need to be pursued. Given the present dearth of suitable wetlands where alligators could potentially be reestablished, much of the attention must be focused on habitat restoration. Ecological restoration is a science unto itself, one that draws on virtually all the disciplines in the field of natural sciences, and a detailed discussion of the requirements for repairing or creating new wetlands habitats is beyond the scope of this book. But restoration ecology and species reintroduction programs are natural partners in the broader field of biodiversity conservation, and collaborations between these two fields will be not only mutually beneficial (Lipsky and Child 2007) but also increasingly necessary.

Ecological restoration is the process of assisting the recovery of a habitat that has been degraded or destroyed. The ultimate goal of restoration is usually to emulate the function, structure, biodiversity, and dynamics of the specified ecosystem. The extent of restoration efforts will depend on the program goals, but it requires a long-term commitment in terms of land and resources, and so must be planned and monitored carefully with as wide a spectrum of stakeholders as possible. For Chinese alligators, a range of options are available, and the goals of each must fit within the often severe limitations of land and resources available for the restoration effort. Most options will likely focus, at least initially, on the creation of habitats that meet the minimal conditions for the establishment of a breeding population of alligators, and not on the reclamation of naturally functioning wetlands. Here we discuss some options for site restoration, beginning first with hill ponds, where land is more readily available but less suitable in biological terms for alligators. We then discuss issues related to the restoration of the more biologically rich valley bottom sites, which while offering a more suitable ecological setting for alligators are under intense land use pressures.

Hill Pond Sites

The Anhui Forestry Bureau's main concerns regarding alligator reintroduction efforts have been the likelihood of conflicts with farmers. As a result, the AFB has focused on trial releases of alligators on lands that they own and manage, not those used by local farming communities. When we first began discussions on the feasibility of releasing captive-reared alligators, the AFB suggested areas in Langxi, Nanling, and Guangde counties on lands managed by local forestry departments. These sites are in low hills dominated by tea plantations or tree farms managed for the production of pine. Because these lands are owned by the county forestry departments, existing ponds could be used or new ones created, and surrounding terrestrial areas could be controlled by the wildlife authorities, thereby minimizing conflicts with local people.

While reintroductions on forestry department lands will avoid potential conflicts with farmers, the habitat is less suitable than in the rich valley bottoms where cultivation is most intense. Much of the rationale for using these areas is the fact that they are not too different from places where some wild alligators are found today, the isolated hill ponds (type 3 habitat in Chapter 7). While this is true, it is simply a reflection of the fact that the more productive lower elevation wetlands preferred by alligators have all been converted into rice paddies. This has forced alligators into a fugitive existence, moving up into ponds in the hills. These areas are usually marginal as alligator habitat, but as a result of the adaptable nature of the alligator, small groups can survive in these sites. And in some of these areas, such as Zhuangtou and Zhongqiao, they have even successfully nested.

In 2001 the AFB made a proposal to the SFA to support the development of six sites for the release of alligators (Table 8.1), all on Forestry Department lands used for production of lumber or tea. The total area of the six sites was 1,327 ha, and the proposal referenced the need for habitat restorations such as connecting ponds, removing water control structures, creating islands, planting a buffer of vegetation around the site, working with local farming communities to reduce the use of pesticides, and ensuring adequate water levels (W. Wang 2001). One issue that was addressed in great detail was the cost, which was considerable. Total expenditures (i.e., salaries, construction, environmental education, monitoring, genetic analyses, and purchase of alligators for release) estimated for the development of these sites over a 5–year period were ¥164 million (ca. $24 million) if the land were purchased and ¥104 million (ca. $15 million) over a 10–year period if the land was leased.

Table 8.1 Proposed alligator release sites

Site	County	Size (ha)
Huagu	Guangde	75.0
Gaojinmiao	Langxi	226.7
Xiadu	Xuangzhou	150.0
Yashan	Nanling	140.0
Sanyuan	Wuhu	68.7
Ganqu	Jinxiang	666.7

Note: In 2001 the Anhui Forestry Bureau proposed developing these six areas as alligator release sites.

The majority of the proposed budget was for purchasing land that in fact is already owned by the AFB, again reflecting the highly commercial nature of alligator conservation efforts. In effect, the provincial government (Anhui) wanted the national

Figure 8.6 (*above and facing page*). Gaojinmiao restoration and alligator release site, Langxi County, Anhui: (a) the ponds under construction in 2003; (b–c) views of two ponds in 2007.

government (SFA) to pay them for the use of their lands to develop the alligator conservation program. The AFB included the purchase of 1,327 ha of farmland, at a price of ¥90,000 per ha (ca. $13,000/ha), for a total land purchase price of ¥1,194,300,300 (ca. $17.5 million). Alternatively, the AFB would rent a hectare of land on an annual basis for ¥4,500, with the total cost over a 10–year period being approximately ¥60 million (ca. $8.8 million). Only 1 of the 6 sites, at Gaojinmiao, was approved and subsequently received support from the SFA and the Anhui government as a trial reintroduction site.

Gaojinmiao, in Langxi County, is a forest preserve set in low hills and has a landscape dominated by tea and pine plantations. The AFB plans to build a network of ponds (ca. 50) that may eventually support a population of more than 100 adult alligators. From 2006 to 2008, 21 alligators were released and monitored using radio telemetry (X. Wu, pers. comm.). All the alligators survived and, with the exception of one individual who moved 2.5 km and settled in a pond in a nearby village, they remained in the newly created wetlands. In addition to creating new ponds, the AFB is also renting nearby areas from farmers and using these, with natural ponds, as alligator habitat. By 2008 they had rented 3.4 ha with eight small ponds. The results of the Gaojinmiao release and monitoring will be important for determining how useful these types of habitats will be as future alligator release sites. If alligators can survive and nest in these areas, as seems likely, a network of hill ponds on AFB lands could support small breeding groups of alligators. In July 2008 an alligator nest was discovered at the Gaojinmiao site (H. Jiang 2008). Nevertheless, the biological productivity of these sites is considerably lower than wetlands in the richer valley bottoms, and the carrying capacity (target density) of alligators at these sites will also be low in comparison.

Valley Bottom Sites

While the forested hills that belong to the Forestry Department offer one option for releasing Chinese alligators, to realistically think about returning alligators to the landscape we must move downhill into the rich valley bottoms that once harbored lakes, ponds, and marshes, but that are now the domain of the farmer. These fertile low-valley sites offer, in a biological sense, the best areas for restoration of wetlands capable of supporting alligators. But they are subject to intensive human agricultural uses. Though these areas are now intensively farmed, the future of land use patterns in the middle and lower Yangtze River valley in the coming decades must also be considered within the context of sweeping socioeconomic changes that are now at work in China. China's unprecedented economic growth over the last two decades has fueled a booming demand for urban labor. Longstanding Mao-era restrictions on movement within the country were relaxed under Deng Xiaoping, and the result has been the largest mass migration the world has ever seen as people flock from rural areas to find work in cities (Yardley 2004). The Chinese government estimates that 114 million rural migrant workers have already found employment in urban areas, and this does not include accompanying family members. This figure is conservatively expected to rise to 300 million by 2020 as urban incomes remain well above those in the countryside. This mass movement is part of a long-term government strategy to shift China from a rural to an urban-based economy. Some of this migration is seasonal, with workers returning to the farms once or twice a year to plant or harvest crops. However, increasingly, families are seeking ways to escape permanently from pervasive rural poverty. At the same time, new advances in agriculture have led to a greater efficiency in farming, particularly with new hybrid strains of rice. Between 1975 and 2000 the amount of land in China devoted to rice production fell from 36.5 to 30.5 million ha, partly due to new hybrid rice technologies that improve yields by 15% to 20% (Xinhua News Agency 2004). The combination of increased urbanization and improved farming efficiency suggests that in the future there may be more opportunities to establish "green" areas in the rural landscape where nature reserves, including restored wildlife communities, can be established. The rich valley bottoms that were alligator habitat, however, are also prime agricultural parcels. Our experiences in the NCAR suggest that communities that own these lands are benefiting from urban work opportunities and leasing their lands to even poorer rural migrants who do not have access to valuable rice-growing lands. While future trends in the use of agricultural lands remain uncertain, it is clear is there will be swift-moving, complex changes in land use practices in the areas suitable for establishing wetlands reserves for alligators and other wildlife. A firm commitment from the Chinese government will be needed to make the best use of the opportunities presented by the changes in rural land use dynamics.

Returning to the present, strategies to improve the existing habitat need to be identified. The few valley bottom sites where Chinese alligators survive today have little or no buffer from farming activities, and conflicts with people are unavoidable. Here, the biggest challenge will be to find ways for local people to benefit from the presence of alligators, and to actively involve the local communities in the safeguarding of their habitat. The logical place to start efforts of this type will be at one or more of the sites where small groups of alligators are found today. The total amount of area at these sites is very small, consisting at present only of the ponds themselves (Chapter 7). As part of programs that include the active participation of farmers, there are options for improving some of these sites. Not all of the designated sites will be suitable for restoration; some are just too small or heavily impacted. Among the best candidates for site restoration are Hongxin, Zhongqiao, Zhuangtou, and Shaungken.

What needs to be done at these sites in order to improve them as alligator habitats? We suggest three principal criteria need to be considered. The first is the supply of water. Many of the ponds used by alligators are also used by farmers as sources of water for growing crops. In some areas, such as Hongxin, irrigation of fields for summer crops results in a significant drawdown leaving alligators exposed (Figure 8.7). This is particularly problematic for juveniles, which are much more vulnerable to wading birds and other predators during periods of low water. Agreements are needed with local communities to ensure that water levels are maintained at minimum levels throughout the spring to autumn activity period of alligators.

The second issue to consider for improving existing alligator sites is the creation of buffer areas. At nearly all of the valley bottom ponds with alligators, the cultivation of rice or other crops extends up to the very edge of the water or beyond, as in many cases the shallow water margins of these ponds have even been converted into rice paddies. There is no natural shoreline vegetation, nor in most cases are there suitable sites for alligators to bask out of the water. The presence of natural shallow water shoreline habitats and a shoreline terrestrial buffer of natural vegetation, even one only a few meters wide, would greatly improve the quality of these sites for alligators. It would also provide a margin of physical separation between alligators and farming activities. Today, in the absence of these shoreline buffers, in many areas alligators have only been able to survive in ponds with small islands, which provide the alligators with a different kind of buffer from farmers. Alligators appear to prefer small islands for nesting and even for construction of their burrows, so creation of islands as part of a habitat restoration program would be recommended even if shoreline buffers were also in place.

The third issue is the physical size of the sites, as most current sites are quite small, rarely more than 2 ha (Chapter 7), and unsuitable for more than a single breeding pair of alligators (Box 8.3). Increasing the size of these sites will require, in most cases, converting agricultural lands to wetlands.

Given the intensive land use pressures on these sites, how can they be improved or enlarged? One approach would be to pay the farming communities to maintain certain minimum conditions or to convert lands currently used in rice production into shallow water habitats around the protected ponds. In January 2004, as a first step toward working with local communities on this issue, WCS and the Anhui Forestry Bureau sponsored a workshop in Jinxiang, one of five counties in the National Chinese Alligator Reserve. The meeting brought together representatives of the Anhui Forestry Bureau, the ARCCAR breeding center, county-level forestry representatives, and leaders from three communities where alligator protected sites are located to discuss options for improving habitat. Participants addressed ways to improve sites by repairing water control structures (broken dams that no longer hold water) and the feasibility of renting lands surrounding the alligator ponds for conversion into shallow water wetlands.

Another strategy would be to identify land uses that are more likely to encourage the coexistence of farmers and alligators and that are able to provide local people with as much, or more, income as do rice farming or other alligator-incompatible activities. This is a subject that will need significant consultation with farming communities, agricultural experts, and alligator specialists in order to identify biologically, economically, and socially suitable shifts in agricultural practices that would improve habitats for alligators and conditions for the farmers.

Alternative land uses and the creation of more "alligator friendly" agricultural landscapes implies above all a switch from rice cultivation, but what are the realistic

Figure 8.7. The Hongxin pond (a) in April 2003 and (b) in August 2003. In August the lake level was drawn down to irrigate fields for the second annual rice crop.

alternatives? We would propose two broad alternatives. One would be to provide farmers with opportunities for alternative livelihoods. This could include microloans and training programs to start businesses in nearby towns or to initiate agricultural activities that produce equivalent incomes from smaller parcels of land (e.g., fish, frog, turtle farming in ponds). Another possibility would be agricultural activities that are more consistent with the maintenance of alligator habitats, such as retaining shallow water margins around ponds for the cultivation of hydrophytes. Cultivation of lotus plants in shallow water areas would be more compatible with alligator conservation, and it produces more than 9 times the income per unit area than the typical double-cropped rice paddy (X. Liu et al. 2004). A shift to seasonal commercial vegetables such as tomato, pepper, and cucumber has also been shown to result in much higher labor productivity that results in 1.5 to 5 times the net income for local farmers (X. Liu et al. 2004). The appropriate choice of new crops, such as fruit trees, or adopting wetland agroforestry could also provide effective buffer areas surrounding the wetlands used by alligators. Cultivation of the poplar (*Populus deltoides*), a short rotation deciduous tree, also allows the intercropping of winter cereal crops at a time of the year when the trees have no leaves, and these types of agroforestry systems have been shown to be 20% to 47% more profitable than crop monocultures (S. Fang et al. 2005).

The integration of wildlife conservation plans with changes in land use practices in rural China will require a new type of synergism between various sectors in the Chinese government. In particular, it will depend on cooperative efforts between the SFA (which manages wildlife at the national level) and provincial wildlife bureaus, and government groups responsible for agricultural production and land use in rural areas. There will also be opportunities for national and international NGOs, particularly as a bridge between local people and the regional and national government, and the development and implementation of new and novel approaches to wildlife conservation.

A third approach to promoting a peaceful coexistence between farmers and alligators would be through programs whereby farmers directly benefit by having alligators in their ponds. Ecotourism is one possibility but is too limited in applicability to be a serious option in most areas. Another alternative would be to allow farmers to sell a certain percentage of the alligators in their ponds each year. This would provide direct economic rewards to communities that invested in habitat improvements for alligators. Given the limited amount of habitat that will be available at most of these sites, as well as the absence of natural predators and dispersal areas, populations of alligators could be expected to grow and at some point exceed the carrying capacity of the available area. This would necessitate the removal of a portion of the population at periodic intervals. A well-managed system for controlling the capture and sale of juvenile alligators would have to be implemented, but this kind of commercially based alligator ranching program would fit in well with plans for large-scale metapopulation management strategy for alligators. The potential for using innovative economic incentives for linking small-scale wetlands reclamation and the creation of breeding populations of alligators within a new type of agricultural landscape is something we feel has great potential.

Larger Wetlands Initiatives

Sites in the NCAR where alligators live are all quite small. Larger wetlands will be able to support larger alligator populations and are more suitable for the restoration of entire communities of wetlands fauna. Ideally, habitat restoration efforts would include areas much larger than those now within the NCAR. We have visited some of the larger wetlands within the historic range of the Chinese alligator and evaluated

Figure 8.8. Potential areas for habitat restoration and reintroduction of alligators.

their potential as sites to restore alligator populations. Some areas, such as the huge Poyang and Dongting lake systems, have sizable potential habitat but are subject to intensive fishing pressure, which is incompatible with alligator reintroductions. There are, however, opportunities in other areas, some of which we have visited and surely many we have not. Some of these sites are existing wetlands with excellent alligator habitat that would require little if any habitat modifications (e.g., Xiazhu Hu). Others are part of larger wetlands reclamation efforts or proposed wetlands parks.

Yancheng Reserve

The Yancheng marshes stretch from latitude 32° to 34° N along the coast of Jiangsu Province, spanning the transition from subtropical to temperate climates and contain the most extensive salt marshes and intertidal mudflats in China. The reserve, totaling 4,553 km², was established in 1983; in 1992 it became both a national nature reserve and a UNESCO Biosphere Reserve. Its coastal wetlands are an important site for a variety of waterbirds and is a vital stopover site along the Asian flyway for some 200 migratory species. Estimates suggest that 1 million waterbirds use the reserve; this number swells to 3 million when migratory species are included. Yancheng is perhaps best known as the home of the largest wintering population (ca. 1,000 individuals) of the red-crowned crane (*Grus japonensis*). The reserve also has a resident population of Chinese water deer (*Hydropotes inermis*) living among the tall grasses of the coastal wetlands. While the area is comprised principally of mudflats, tidal creeks, and salt marshes, there are extensive areas of freshwater marsh that have been created by the diversion of water from several rivers. It is some of these areas that may have potential as alligator release sites.

The reserve is zoned into core, buffer, and experimental areas. Although the latter two sectors are considered part of the reserve, the reserve management has no legal rights to these areas. They are owned by local communities who use them for agriculture and aquaculture. The core region of the reserve covers 17,400 ha. Parts of the core area are used as fish ponds or diked and flooded, ostensibly as a means to improve habitat for wading birds, including the cranes. Adjacent to Yancheng is the Dafeng National Nature Reserve. Covering an area of 40 km², it is administered by the SFA principally for the establishment of a herd of Père David deer (*Cervus davidanus*) (Box 8.5).

One of the principal questions about the suitability of the Yancheng Reserve as a potential site for the release of Chinese alligators is whether alligators would be able to survive the cold winter temperatures that far north. While Yancheng is a coastal region buffered from the temperature extremes found farther inland at this latitude, temperatures in the coldest months (January–February) can reach freezing and temperatures of –10°C have been recorded. Alligators would be hibernating in burrows below the surface during the winter, but the effects of these low temperature are unpredictable. Chinese alligators in their present area of distribution appear to be at or near the northernmost limit of their temperature tolerance (Chapter 3), but this site may become more climatically favorable as global warming progresses.

Chongming Island

Chongming is a large alluvial island (76 × 15 km) at the mouth of the Yangtze River, within the municipality of Shanghai and immediately to the north of the city. Chongming Island was first mentioned in about AD 600 as a small sandbar. By 1200 it had become a large island with permanent villages (Van Slyke 1988). Today, Chongming Island is the world's largest alluvial island, and it is growing rapidly (at least prior to the closure of the Three Gorges Dam) with some 5 km² of mudflats accreted onto the downstream, eastern tip of the island each year. These extensive mudflats are used by many resident and migratory bird species. Like Yancheng it is one of the most important stopover sites for migratory birds on the East Asia–Australasian flyway, with 312 species of birds recorded. In November 1998 the Shanghai municipal government created the Chongming Dongtan Nature Reserve, covering 326 km² of the eastern end of the island. Besides protecting the intertidal mudflats and marshes, the goals of the reserve are to promote environmental education and ecotourism. In 2002 the site was designated a Ramsar wetland of international importance and in 2005 the mudflats habitats were declared a national migratory bird reserve.

The extensive tidal mudflats that characterize Chongming Dongtan are not suitable for alligators, but nontidal freshwater habitats have been created by periodically building levees along the eastern end of the island. Since the early 1950s approximately 500 km² of land have been altered in this fashion by the Shanghai municipal government for agriculture and aquaculture. In 1998 a significant portion of the reserve core area was diked to create freshwater ponds for fish culture; some of this land is now being developed, in three phases, as a wetland park. The first phase was a small trial restoration of a 10-ha area that will serve as the visitor entrance. Phase 2, a 120-ha region surrounding the initial phase, was completed in 2007. The full area will cover 860 ha and will offer extensive habitat for native Chinese wetlands fauna. The site is dominated by a variety of vegetation characteristic of freshwater wetlands and has an abundance of potential prey items for Chinese alligators, including crabs (*Eriocheir sinensis, Metaplax longipes*), aquatic insects, snails (*Bullacta exarata*), mollusks (*Moerella iribescens, Sinonovacula constricta*), shrimp (*Macrobrachium nipponense*), and a variety of frogs and snakes.

In June 2007, a trial release of 6 Chinese alligators (3 males, 3 females) took place in the restored wetlands of Chongming. The 6 alligators were chosen based on provenance (3 from the Changxing breeding center in Zhejiang Province and 3 U.S.-born animals) and were released following health screening by a qualified team of veterinarians. The U.S.-born animals were shipped to China in May 2006 and had spent one year in quarantine at the Changxing center. All animals were fitted with radio transmitters and followed by a team from East China Normal University and the Shanghai Industrial Investment Corporation, which manages the wetland park. Two animals died shortly after release when they drowned in fishing nets hidden underwater in canals adjacent to the release site. An attempt was made to remove all fishing nets, and the remaining animals have adapted well. The first alligator nest was laid in 2008 and successfully produced 12 hatchlings. Hopefully, this trial effort will be expanded; Chongming has the potential to support a significant population of alligators.

Xiazhu Hu

The Xiazhu Hu wetland, in Zhejiang Province, offers an outstanding opportunity for the establishment of a sizable breeding population of Chinese alligators in a comparatively natural wetland system. Xiazhu Hu is a complex of shallow-water lakes (1.5–2.8 m deep) and canals in Deqing County within the drainage of the Tai Hu lake system. The area was used extensively for the cultivation of hydrophytes (particularly water chestnut, *Trapa natans*) and fishing, but is in the process of being established as a provincial wetland park under the Zhejiang Tourist Bureau.

We visited the site in 2005 and 2006 and were impressed with the quality of habitat and the potential for this site as a wetland where alligators and other native aquatic fauna can be reestablished. One feature of the site that makes it particularly suitable for alligators is the presence of more than 600 small islands that provide an ideal matrix of wetland and terrestrial habitats. Although the wetland reserve is quite large (3,600 ha), only the core area of 1,150 ha (Q. Wang 2006) is presently suitable as a wildlife reserve. This is by far the largest potential reintroduction site that has been identified to date, and the incorporation of Xiazhu Hu into a national plan for the restoration of wild populations of Chinese alligators would be perhaps the most important single step that could be accomplished at this point.

RELEASE THE DRAGON: ALLIGATOR REINTRODUCTIONS

As a conservation tool, captive breeding is a stopgap measure to prevent the extinction of species whose wild populations have a high probability of disappearing. Captive breeding can play a key role in reintroduction programs designed to reestablish viable, free-ranging populations of animals that have been extirpated from particular areas or that are extinct in the wild. Growing human population pressures and shrinking habitats place increasingly severe strains on wildlife, and for an increasing number of species, habitat restoration and links between captive and wild populations

BOX 8.3

How Much Habitat?

A key question in planning alligator conservation efforts is, how much land do they need? Obviously, the more habitat available for alligator reintroductions the better, but for planning purposes we need an idea of what range of densities we might expect for these populations. How do we estimate the target carrying capacity for a selected area? Alligators require a mixture of aquatic and terrestrial habitats (Chapter 5), including a variety of wetland types for different size classes of animals (juveniles, adults), and terrestrial spaces for nesting, burrowing, and basking. The amount of habitat required by a population of Chinese alligators depends on many variables, including the structural complexity of the site (e.g., a series of small ponds, each with associated marsh and islands versus one larger pond with no islands) and its biological productivity (i.e., food availability).

Observations of existing alligator populations, as well as experience with the American alligator, suggest that the presence of small islands as well as irregular shorelines greatly increases the suitability of habitats for alligators. Here we offer some ballpark estimates of the habitat requirements of Chinese alligators. The objective of these calculations is to estimate what may be an appropriate density for a natural, self-sustaining population in order to provide preliminary guidelines for establishing new populations by reintroduction. Follow-up studies of the released alligators will be required to determine more precisely how large a population of alligators can be supported at each site.

Given the limited base of information on wild populations of Chinese alligators (Chapter 5), we derive an estimate of habitat requirements based on American alligators. For the purpose of our calculations, an appropriate measure, or index, of abundance is the density of nests in a given area. A study of American alligators in Florida (Woodward et al. 1992) found an average density of 1.15 nests per ha of suitable nesting habitat. In four areas in Florida, suitable nesting habitat comprised 1% to 1.5% of the total habitat available for American alligators. For our calculations we make two important assumptions, the first is that because habitats for Chinese alligators will be restored, we presume that habitat improvements can result in up to 10% of the total area being suitable for nesting. We also assume that as a result of the much smaller size of the Chinese alligator, average nest density is higher than for American alligators; we use a figure of 2 nests per ha of suitable habitat. Thus Chinese alligators would need 5 ha of habitat for every nest. But adult female alligators do not nest every year; for American alligators it is usually less than 50% of the female population, although this figure is likely density dependent. Based on a 50% nesting rate for adult female Chinese alligators, a 5-ha site would be suitable for 2 adult females (each one nesting every other year). The amount of habitat needed for 1 female (and 1 male) would be 2.5 ha. This is a very rough estimate, and we emphasize that it is only a general guideline for planning trial reintroductions and initial target densities.

will assume a critical role in conservation programs. For a variety of species, reintroduction programs are playing an increasingly important role in many parts of the world (Fischer and Lindenmayer 2000). Nonetheless, the release of captive-reared animals is always risky and has had a mixed history, with many programs failing for a variety of reasons (Beck et al. 1994). The release of captive-bred mammals and birds back into the wild can be an expensive and time-consuming process (Miller et al. 1996; T. White et al. 2005). Survival traits of these animals often have to be learned, which may necessitate extensive periods of training. As the number of attempts has increased, reintroduction has slowly evolved into a more proven methodology (Seddon et al. 2007). Attempts tend to be more successful if they are conducted initially on an experimental basis and carefully monitored (Fischer and Lindenmayer 2000; Stanley Price and Soorae 2003). As a result of the increasing occurrence of reintroduction projects worldwide and the growing need for specific policy guidelines to help ensure that the reintroductions achieve their intended conservation benefit, the IUCN-SSC established a Reintroduction Specialist Group in 1988 that has produced

a series of guidelines for maximizing the chances of success of reintroduction efforts
(IUCN 1998) (Table 8.2). The linking of captive and wild populations of Chinese
alligators is treated in the section on metapopulation management. Here, we discuss
the potential for using captive-bred alligators as a source population to restore the
species in the wild.

Reintroductions of Chinese Alligators

Chinese alligators have a number of advantages as a candidate species for reintroductions. Aside from the fact that they are extremely adaptable and resilient, alligators
are reptiles that in general are relatively easy to rear in captivity and do not need to
learn survival skills in order to successfully adapt to wild habitats. Broadly speak-
ing, the reintroduction of endangered species of reptiles is simpler and less costly
than it is for mammals and birds. Preliminary evidence also suggests that reintroductions of reptiles and amphibians are on average more successful than those for
birds and mammals (Beck et al. 1994). The fact that they are ectotherms, with much
lower metabolic rates than birds and mammals, may also be an advantage (Box 8.2).
Although the conservation value of reintroduction, or population augmentation,
efforts for reptiles has been debated (Dodd and Seigel, 1991; Burke 1991), this type
of management has been relatively successful for the Crocodylia (Dodd and Seigel
1991). The release of captive reared individuals to speed the recovery of critically
endangered wild populations has been used as a management tool for crocodilians

Table 8.2   IUCN guidelines for wildlife reintroductions

Prerelease activities:

Selection of release sites. Sites should be within the historic range of the species and evaluated for suitability. The causes of the previous decline of the species should be identified and either reduced or eliminated.

Choice of stock. Animals should preferably come from a wild population that is closely related genetically to the original native stock and show similar ecological characteristics to the original subpopulation. If captive-reared animals are to be used they must have been managed both genetically and demographically.

Veterinary screening. Animals to be released should have health and genetic screening.

Socioeconomic studies and political support. Socioeconomic evaluations should be made of the impacts, costs, and benefits of the reintroduction program. Programs are generally long-term and require considerable economic and financial support.

Release strategy. Develop a plan for the number, size, sex, and timing of release for animals at each site.

Conservation education. Establish educational and public relations programs for local communities and a wider audience; involve local people in the program; provide professional training for individuals.

Indicators. Identify short- and long-term indicators of success.

Monitoring. Carefully design pre- and postrelease monitoring to evaluate the success of the program.

Postrelease activities:

Monitoring. Monitor demographic, ecological, and behavioral studies of released animals and their long-term adaptation.

Program review. Evaluate the need for program revision.

---

in a range of countries, most notably in India, Nepal, and Venezuela. Nevertheless, reintroductions, including those for crocodilians, are doomed to failure unless the underlying causes that resulted in the species' threatened status are addressed. This has been aptly demonstrated by the conservation woes of the Indian gharial (*Gavialis gangeticus*) (Gharial Working Group 2007).

Some attempts were made by the AFB to release Chinese alligators prior to the start of our work, but these trials were carried out in an unsystematic fashion and there was no follow-up or monitoring. Y. Zhou (1997) mentions the release of 150 captive-reared alligators in 1990, and Z. Wan et al. (1998) report that a total of 69 were released between 1987 and 1997 (9 in 1987, 15 in 1989, 15 in 1990, and 30 in 1997). No published details on these releases are available. In 1997 the AFB reported releasing 2 captive-reared alligators at Yanlin, one of the 13 official sites in the reserve. That program was deemed a failure because the alligators reportedly did not adapt well and there was local opposition from farmers. The AFB has routinely translocated alligators when they are found by farmers in an inconvenient location (such as a rice paddy). These animals are captured and released by the AFB at one of the sites designated for the protection of alligators. No real effort has been made by the Anhui Forestry Bureau to follow up on the results of these translocations. Nevertheless, it is apparent that the success rate has been very low and animals usually do not remain very long at the release site. Accidental releases have come in the form of escapes from the ARCCAR. In 1991 flooding allowed approximately 30 three- to four-year-old alligators to escape (J. Fu 1994). During our visit, we saw one alligator living in a pond just outside the uppermost of the breeding enclosures.

Although alligator releases, both planned and accidental, have occurred, they have not been part of any overall strategy for the maintenance of viable wild populations. The rationale for the establishment of the breeding centers in China in 1979 was to prevent the species from going extinct, but no explicit reference to the use of captive animals for reintroductions was given in the early descriptions of the program or in the 1992 proposal to CITES. An evaluation of the Chinese program carried out by the IUCN Crocodile Specialist Group did not place much stock in the idea of reintroductions given land use pressures in the area (Webb and Vernon 1992), and a systematic program of reintroductions to reestablish breeding groups in the wild was not considered at all by the Chinese authorities in Anhui or Beijing prior to the initiation of our work in 1997. Today, the Chinese government has expressed more interest in reintroductions (H. Jiang et al. 2006), but only limited steps have been taken to implement this policy. Nevertheless, as of the spring of 2007, three small-scale trial releases of captive-reared animals had taken place.

The first release of captive-reared alligators was a joint effort conducted by East China Normal University, the Wildlife Conservation Society, and the Anhui Forestry Bureau in 2003. This was the first time that an experimental release was carried out in a planned and systematic fashion, with extensive follow-up monitoring using radiotelemetry. Three young adult alligators (2 females, 1 male) from the ARCCAR were fitted with radio transmitters and released at the Hongxin designated site at the end of April 2003 (Ding et al. 2004) (Figure 8.9). This effort was deemed a success as all 3 individuals survived and the following year 1 of the females successfully nested. At least 1 of the 2 females has nested in each subsequent year (2005–2008)

The second experimental release, in 2006, was at the Gaojinmiao site in Anhui

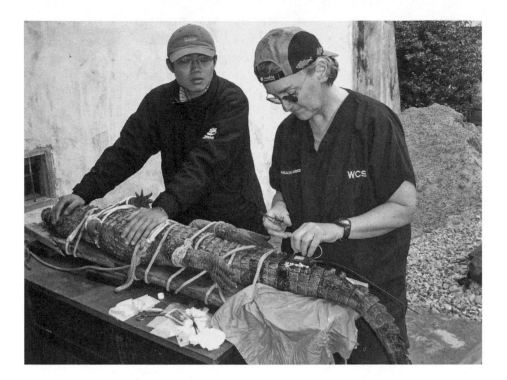

Figure 8.9. A radio transmitter is attached to the tail of an adult male Chinese alligator at the ARCCAR by Bonnie Raphael, Wildlife Conservation Society senior veterinarian, assisted by Wang Zhenghuan, ECNU graduate student, April 2003.

Province, on lands owned by the Langxi County Forestry Bureau. With support from the AFB, six ponds were created to offer wetland habitat for small groups of alligators. Because the land is owned by the forestry office there is no need to work with local communities. Six alligators were released in Gaojinmiao in April 2006. They were initially monitored by Wu Xiaobing and students from Anhui Normal University using radiotelemetry. Because the radios did not remain fixed to the alligator's tails, the chief monitoring technique became night counts (X. Wu, pers. comm.). All 6 alligators survived their first year in the wild. A second release of 6 alligators at this site took place in May 2007, and another 9 were introduced in 2008 (H. Jiang 2008). Nesting at Gaojinmiao was reported for the first time in 2008 (H. Jiang 2008).

The third release site for Chinese alligators was in the newly created wetland park on the eastern end of Chongming Island, in Shanghai municipality. In June 2007 6 alligators—3 from the Changxing Breeding center and 3 U.S.-born Chinese alligators from the AZA species survival plan—were released as part of a joint program between the Shanghai government, the Changxing breeding center, WCS, the Saint Augustine Alligator Farm and Zoological Park, and the Walt Disney Company. Unfortunately, within two months of the release 2 of the alligators had drowned in underwater fishing nets in canals adjacent to the release area. Efforts to remove nets were redoubled and discussions held with representatives of the local communities. The problem now appears to have been resolved. The remaining 4 alligators successfully hibernated and were monitored in the spring of 2008. Successful nesting by one of the females occurred in 2008. The results of these experimental releases demonstrate that captive-reared Chinese alligators adapt well following release into sites outside the breeding center. They excavated burrows for overwintering and successfully reproduced at all of the release sites, a not surprising result given the resilient and adaptable nature of the species. This behavioral and ecological plasticity is a good indication that alligators will be able to adapt to a wide variety of wetland types within the species' historical range.

Given the large number of alligators in captivity, reintroductions offer a significant opportunity for the restoration of wild populations, provided that limitations of habitat availability and government support for these programs can be resolved. Nevertheless, much work remains to be done to develop a science-based reintroduction program. Of particular interest is the issue of the size of habitat patches. How much area is needed to hold a population of a given size? Another way to look at this question is, given a restored area of a certain size, how do we go about establishing target densities for the alligator population? What habitat features are most important in planning habitat restoration? What are the best sex ratios to use when releasing groups of alligators? Is it preferable to release adult, juvenile, or a mixture of age classes? Answering most of these questions will require an adaptive management approach involving a series of trial releases followed by monitoring of established success criteria and the incorporation of lessons learned into modified management protocols.

A CONSERVATION METAPOPULATION: LINKING CAPTIVE AND WILD GROUPS

The special problems inherent in the conservation of small populations have been the focus of much of the theoretical work in the field of conservation biology (Caughley 1994). Together with the longstanding work by ecologists on the regulation of natural populations, this has led to interest in the use of metapopulation analysis in the design of conservation strategies for some threatened species (Hanski and Gilpin 2008). A metapopulation is a group of populations distributed among patches of suitable habitat that are linked by occasional dispersal; it is a dynamic model of how natural populations are believed to exist. Individual subpopulations grow and shrink as a result of a variety of factors. In some of these subpopulations, the number of animals may become so small that they become locally extinct. Depending on a series of factors, such as habitat patch size or habitat quality, local extinctions may be more likely to occur in some patches than others. But as a whole the metapopulation survives as a result of dispersal and recolonization events.

The concept of a metapopulation was originally developed to describe the population dynamics of species that are distributed in discreet habitat patches, such as insect pests in agricultural fields (Levins 1969), and because suitable habitats are patchy by nature, most species are thought to exist as a metapopulation in the wild. Associated with the concept of metapopulations is the "source-sink" model of population dynamics, which recognizes that some habitats are more suitable for a species than others and that by benefit of higher rates of reproduction, survival, or both, these areas serve as a source of individuals for less suitable habitats (sinks). Some populations may be prone to extinction for reasons other than just small habitat patch size or habitat quality, including adverse human impacts such as hunting.

BOX 8.6

China's Plan for Conservation and Reintroduction of Wild Chinese Alligators: What Is the Commitment?

In late August 2001 a group of Chinese and representatives from other countries gathered in Hefei, the capital of Anhui Province, to develop a blueprint for the conservation of the Chinese alligator. After three days of productive discussion, the attendees drafted a document that acknowledged the success of the captive breeding programs and outlined the need to place considerably more emphasis on the recovery and restoration of wild populations. The final document was published in the 2001 Proceedings of the International Workshop on Conservation and Reintroduction of Chinese Alligators. The main long-term goals were

– improve habitat conditions in current groups of Chinese alligators with the objective of establishing viable populations;
– establish viable populations of Chinese alligators in 15 to 20 suitable wetlands in the middle and lower reaches of the Yangtze River;
– seek ways to integrate human activities compatible with the coexistence with alligators, reduce conflicts, and contribute to the local environment and socioeconomic development.

The document also presented a 10-year plan that outlined work toward two main objectives: the conservation of the remaining wild populations and the creation of new wild populations by releasing captive-bred alligators. Plans for the conservation of the remaining wild populations listed the following objectives:

– develop a strategic action plan for the conservation of existing groups of wild Chinese alligators;
– enlarge and restore wetlands currently occupied by wild groups that have the potential to be sustained;
– resolve land-tenure conflicts at protected sites;
– reverse negative population trend of wild alligators;
– establish effective participatory management practices at sites with wild alligators; and
– develop appropriate research and monitoring programs.

For the reintroduction of captive-bred alligators to establish new breeding populations, a similar set of objectives was defined:

– develop a strategic action plan for the conservation of alligators placed in the wild;
– carry out trial reintroductions to define criteria for suitable habitats and to improve guidelines for release strategies;
– gradually reintroduce captive-bred individuals to selected sites to establish new viable wild populations;
– establish protection, management, and monitoring protocols for reintroduced populations; and
– promote integrated development of conservation and local economies.

Although the workshop recognized the need for a significant shift in thinking about how to approach conservation of Chinese alligators, with a greater emphasis on wild populations, the short-term results of the workshop were just the opposite. In the years following the Anhui meeting, the SFA, the Anhui and Zhejiang provincial governments, and local communities invested millions of dollars in improvements in the two principal captive breeding centers.

Following the 2001 meeting the SFA approved ¥15 million (ca. $2 million) for alligator conservation and reintroduction in Anhui Province. Of this total, ¥9 million was supplied by the SFA and the remainder from local government cofinancing. The SFA support was divided into ¥5 million for the captive breeding and reintroductions and ¥4 million for public education. The public education funds have not yet materialized. and the ¥5 million supplied as of 2006 went almost entirely toward improvements in the Xuancheng captive breeding facility, not into reintroductions. The only funds applied toward reintroduction efforts went to the construction of new ponds at the Gaojinmiao site. Nothing was done to improve the sites where wild alligators live. While the release of 21 animals by 2008 in the Gaojinmiao project indicates some commitment toward the goals outlined in the Hefei 2001 document, the funding priority for the Chinese government is clearly directed at supporting captive breeding programs.

Techniques for the management of small populations as metapopulations have been developed for endangered species breeding programs in zoos around the world (de Boer 1994). In North American zoos, under the guidance of the Association of Zoos and Aquariums, SSPs manage breeding of captive groups to maintain a healthy and self-sustaining population that is both genetically diverse and demographically stable. Breeding programs developed by zoos to maximize the survival probabilities of their captive populations entail, in essence, the creation of a metapopulation that links the member breeding institutions. Instead of animals naturally dispersing from one habitat patch to another, animals are transported from zoo to zoo for breeding according to a master plan developed to minimize loss of genetic diversity. As differences between captive and wild populations are increasingly blurred, the need to link them grows, and the approaches developed for zoos can be extended to include a combination of captive and small wild populations. In fact, the metapopulation concept has already been extended to planning of conservation management strategies, including the consideration of captive groups as local populations (Craig 1994) and the linking of captive and wild groups for certain animals, such as the endangered South American lion tamarins (Kleiman and Rylands 2002). We believe this approach holds considerable potential for the Chinese alligator. Indeed, the development of a "conservation metapopulation" that links the various breeding groups with existing and future wild groups will be essential.

By its very nature, this type of a program will require strong central support and guidance to link the efforts of the various groups, spread across the provinces that are involved in Chinese alligator programs. A master plan for the future of the Chinese alligator, based on genetics, demographics of the captive and wild populations, and the restoration and management of habitat, needs to be designed and put into effect. Such a plan should draw on the expertise of a large body of professionals and a host of stakeholders. In the rest of this section, we discuss what we envision as a way to address the complex and multidisciplinary problems that will emerge in an attempt to fully integrate the captive breeding populations into a comprehensive conservation program for the species. The overall coordination of the development and implementation of the master plan guiding this work is something that, by necessity, will need to be carried out through the SFA at the national level.

Genetic Factors

The Chinese alligator, with a total of approximately 150 wild animals, divided into subgroups with a maximum of 2 adult females, represents the epitome of a small, highly fragmented population. The tiny remnants of wild groups are at the mercy of a gamut of factors—genetic, demographic, and environmental—that can lead to their ultimate extermination. Understanding these factors, and how they affect local populations (wild and captive) as well as the entire metapopulation, is critical. Genetic variation is the raw material that permits evolutionary adaptation. In small, closed populations (without immigration), the depletion of genetic variation is an inevitable consequence of the low number of adult animals breeding. In the case of species that are heavily reliant on the existence of a population of captive individuals, it is important to consider critical parameters of the captive population other than its total size: the number of founder individuals and the "effective" population size.

An essential component of an overall conservation plan will be to ensure that genetic variability is not further eroded in the captive populations and that is represented to as large a degree as possible in the reestablished wild groups. The genetic diversity of the population of captive Chinese alligators is a function of the initial number of founder animals used to establish the breeding centers and the genetic diversity among them. The establishment of a new population from a relatively small number of founders will likely lead to a loss of genetic variability and a dramatic alteration in allele frequency (Allendorf and Luikart 2007). If the founders are not representative of the greater population, the captive population will be a biased representation of the available gene pool. This does not appear to be the case for the Chinese alligator as the captive population at the ARCCAR is derived from a relatively large sample (212 wild captured alligators). However only 76 of these founder alligators are believed to have successfully reproduced (X. Wang et al. 2006). The much smaller Changxing breeding center is derived from a founder population of 10 individuals (J. Xu et al. 2004). The total number of founders for both the ARCCAR and the Changxing breeding centers is in excess of 220 individuals, with more than 80 having reproduced. They are likely a large sample of what was an already severely reduced wild alligator population in the 1980s. To some extent the genetic diversity of the ARCCAR captive population has also been augmented through the opportunistic collection of eggs from wild nests, a practice that continued through 2002. Nevertheless, genetic evaluations of the ARCCAR population, and the much smaller group from Changxing, found very low levels of genetic variability when analyzing the nuclear DNA using randomly amplified polymorphic DNA (Wu et al. 2002) in the D-loop region of the mitochondrial DNA (Y. Wang et al. 2003) and based on ALFP analyses (Y. Wang et al. 2006). Work using microsatellites has confirmed low genetic variability of the Changxing population; when compared to American alligators, *A. sinensis* were found to have an average of 46% fewer alleles over a sample of 16 loci (J. Xu et al. 2004). While there have been no genetic evaluations of the small groups of wild alligators, it is known that the survival of some of these groups is the result of the successful reproduction of a single female during recent years, suggesting a high degree of relatedness among alligators at individual sites.

Somewhat surprisingly, mitochondrial DNA genetic variability is also low in the American alligator (Glenn et al. 2002). While American alligators passed through a moderate population bottleneck in the nineteenth and twentieth centuries when numbers were reduced to relatively low levels from commercial hunting, this is not believed to have had much of an effect on genetic variability of the species. Instead, low levels of variability in the mitochondrial DNA of *A. mississippiensis* have been interpreted as consistent with a population expansion following the late Pleistocene when climatic conditions were presumably unfavorable (Glenn et al. 2002). Chinese alligators may have experienced a population bottleneck similar to that of the American alligator in the late Pleistocene but also have been subjected to an extreme reduction in population size as a result of the transformation of their habitat over the last 1,000–plus years. Consequently, the even lower levels of genetic variability found in *A. sinensis* are not unexpected (Glenn et al. 2002).

As a result of genetic drift, the random fluctuation of gene frequencies, populations are projected to lose a certain percentage of their genetic variation every generation. While it is true that substantial genetic changes are more likely in small populations than in larger ones, when considering the likelihood of genetic changes in a population such as that of the Chinese alligator, it is not the total number of individuals but rather the "effective" population size that matters. The effective population size (= N_e) is smaller than the total population (= N) because not every individual has the same probability of passing its genes on to the next generation. Factors that act to decrease N_e and accelerate the loss of genetic diversity tend to be those that act to skew or increase the variance in sex ratio, the number of breeders, or offspring produced. Among crocodilians, where dominant males may monopolize breeding opportunities, N_e will be reduced accordingly. Although N_e is a central tenet in conservation biology, empirical estimates have been difficult to obtain directly from natural populations because key demographic parameters are not available throughout the entire life span of individuals. For most species, values of N^e are thought to be 10% to 20% of the census population size among captive and wild populations (Frankham 1995).

Genetic management of the Chinese alligator metapopulation must make the best use possible of the available stock of animals in captivity. The release of captive-reared animals of compromised genetic variability is known to have deleterious effects on wild populations (McGinnity et al. 2003). In extreme circumstances such as that of the Chinese alligator, the probability of mating between close relatives increases dramatically, resulting in inbreeding depression. As a result of the reduction in genetic heterozygosity, it is more likely that recessive, deleterious alleles may be expressed, potentially reducing fertility, increasing mortality, or both. A study by J. Xu et al. (2004) gave some cause for optimism for future conservation efforts as it found moderate levels of heterozygosity and the presence of rare alleles in the Changxing population. Future work should focus on the genetic screening of all adult animals in the ARCCAR and Changxing and the development of a breeding program that maximizes the genetic diversity of the progeny. Among the relatively small sample (37) of animals from the ARCCAR sampled for genetic testing, 3 individual alligators were found to have more than half of all the observed polymorphic loci. With more complete screening much, if not all, of the remaining genetic variability of the captive population can be maintained.

Another important factor for conservation programs that rely heavily on captive breeding is the danger that captive conditions can lead to changes in the phenotype of the animals being reared. These can occur in individuals (without genetic changes) or over a period of generations as a result of relaxed selection for traits important for reproduction and survival in the wild (Gilligan and Frankham 2003). Behavioral changes are more likely in other taxonomic groups, such as mammals, but very little is known about this for reptiles. Because alligators have relatively long generational times, genetic changes may also be slow to manifest, particularly as animals in the two main breeding centers are (as adults) maintained under seminatural conditions.

Demographic Factors

The demographic characteristics of a population, including reproductive success, mortality schedules, and sex ratio, can change significantly over relatively brief periods of time. Large populations are buffered from these variations, a luxury not available to smaller groups. Chance variations in the numbers of individuals or key demographic characters can quickly doom a small population. With such small populations, demographic stochasticity also looms as an important factor shaping population viability and may worsen an already gloomy outlook for survival. One prime candidate for potentially deleterious demographic events among crocodilians concerns sex ratio. The sex of Chinese alligators, and all other crocodilians, is determined by the incubation temperature of the eggs (Chapter 5), and nests normally produce skewed sex ratios of either predominantly males or females (Lang and Andrews 1994). The temperature at which eggs are incubated, in turn, is influenced by a number of environmental factors, including the location of nests and materials used to build the nest mound. In general, crocodilians tend to be creatures of habit when it comes to nesting, frequently using the same location on a repeated basis. This appears to be the case with the Chinese alligator. Furthermore, environmental modifications around the ponds likely limit the options for nest sites available to adult females. As a result, a population with a single female nesting over a period of years may produce offspring that, aside from being siblings, may be highly biased toward one sex or the other, which over time could spell demographic doom.

Environmental Variability and Catastrophes

The most obvious environmental factor impinging on the survival of the Chinese alligator is the lack of natural habitat. Exacerbating this is the increased fluctuation in environmental extremes resulting from anthropomorphic alterations of the environment. The habitat in which alligators find themselves today is largely a landscape denuded of its natural ecosystems and converted to agriculture (Chapter 7). Given the lack of environmental buffering of the areas where alligators remain, they are highly vulnerable to events such as flooding and drought. Furthermore, the frequency and intensity of these events have increased because of the wholesale environmental degradation in the region and have played a significant role in the demise of the last populations of Chinese alligators (Chapter 6). For instance, even relatively minor floods can eliminate the entire recruitment for one year by destroying nests. Catastrophic events such as major floods resulting from the breaches of levees are also know to have washed away or killed entire groups of alligators. Another catastrophic scenario is the introduction of a virulent disease. This is of particular concern for the large groups of captive animals.

A metapopulation approach for the management of both wild and captive populations would be the most effective way to plan a habitat restoration and alligator reintroduction program and minimize the genetic, demographic, and environmental risks inherent in working with a small and highly fragmented population. For the immediate future the largest subpopulations of alligators in this conservation metapopulation would be in the two main breeding centers (Figure 8.10). More complete genetic screening of captive animals, combined with individual marking and dividing the current large breeding groups into smaller units, will facilitate pedigree analysis and allow the managers at the ARCCAR and Changxing to keep track of the degree of relatedness among individuals. For captive-managed species with known pedigrees, individual animals can be selected to create reintroduced populations with specific genetic compositions. Based on a master plan built on genetic and demographic information of animals in the metapopulation, and the individual characteristics of each of the subpopulations, animals would be moved from one subpopulation to

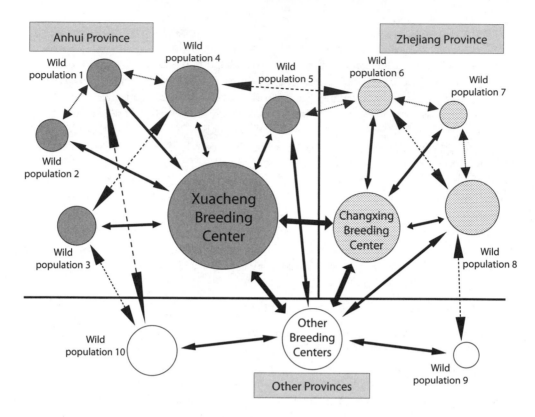

Figure 8.10. Schematic of a Chinese alligator conservation metapopulation. Circle size is roughly proportional to the number of animals in each subpopulation. Thick solid arrows represent exchange of animals between breeding centers; thin solid arrows indicate releases of captive animals and the movement of wild-born animals into the breeding centers. Dotted arrows indicate translocations of alligators between wild populations.

another. Alligators would move between the breeding centers and the wild populations (reintroductions), but as wild populations grow there would likely be a need to remove a specific fraction of animals on an annual basis to maintain target densities. This offers potential for linking alligator conservation to economic gains for local communities by paying them for alligators that are brought back to the breeding centers or translocated to other wild sites. Monitoring of the wild populations nesting, combined with periodic genetic testing, would allow wild-born animals to remain part of the metapopulation pedigree analysis.

THE CONSERVATION CHALLENGE: SETTING RECOVERY GOALS

In the process of developing recovery plans, managers of threatened or endangered species must establish recovery criteria and goals for management of critical populations. What should the recovery goals be for the Chinese alligators? In the past, the foremost goal in China has been to create a large captive population, thereby diminishing the possibility that the species will go extinct. But as Chinese alligator conservation enters a new phase, that of the recovery of wild populations, specific attainable goals need to be identified. At the 2001 conference on Chinese alligator conservation in Hefei, Anhui Province, there was a great deal of discussion about the potential for the restoration of wild alligator populations. One specific proposal for Anhui Province included benchmarks for habitat restoration and reintroduction (W. Wang 2001). The proposal was for the establishment of 6 sites, totaling 1,327 ha, each able to maintain a wild population of 300 to 500 alligators. While there were no details on how target population densities were estimated, the overall target population for wild alligators under this plan would be 1,800 to 3,000 animals. At a national level, the Hefei workshop also produced a first attempt at an overall set of goals, outlining the importance of habitat issues and the stated goal of establishing 15 to 20 new wild alligator populations over the next 10 years (Box 8.6). Unfortunately, this proposal has received no backing from the SFA.

Here, we present what we believe to be a realistic set of objectives for returning Chinese alligators to the wild. Our proposal is twofold, one a relatively short-term goal, but perhaps with great symbolic significance, the other a final goal that will signify recovery:

> Within a 15-year period, establish sufficient wild alligator populations to achieve a downlisting, based on IUCN criteria, from its current level of critically endangered to endangered.
> Within a period of 30 years, have a quantifiably viable wild population of Chinese alligators.

Downlisting the Chinese Alligator to Endangered

The IUCN Red List of threatened animals grew out of efforts in the 1960s by Peter Scott to draw public awareness to the issue of biodiversity losses and has matured into a worldwide effort coordinated by the IUCN–World Conservation Union and its network of specialist groups. Since the 1980s, the often ad hoc method for classifying species according to their threatened status has developed into a quantitative system to evaluate the relative risk of extinction (IUCN 2001). Currently the IUCN lists the Chinese alligator as "critically endangered," a well-deserved classification given the status of its wild populations. There is a very real danger that in the future it could be uplisted one notch to "extinct in the wild." We believe, however, that given sufficient backing by the Chinese government, the species can actually head in the other direction and be downlisted to endangered within a period of 15 years. This would be a remarkable achievement from the point of view of the species status and the change in category, from critically endangered to endangered, would also be an international recognition of China's wildlife conservation efforts. Here we outline how this might be accomplished.

The IUCN has established quantitative criteria to classify the threat of extinction to species; these criteria are used to place the species in its corresponding Red List category (Box 8.7). In order to be removed from the critically endangered (CR) category, it is necessary to demonstrate that a species does not meet any of the five criteria (A–E) and that it has not done so for at least five years. The five criteria are (A) declining population (past, present, and/or projected), (B) geographic range size, and fragmentation, decline, or fluctuations, (C) small population size, and fragmentation, decline, or fluctuations, (D) very small population or very restricted distribution, and (E) quantitative analysis of extinction risk.

Under criterion A, for the alligator to no longer qualify as CR, the population must have declined less than 80% over a period of 3 generations. A generation is defined as the average age of parents of the current cohort (i.e., newborn individuals in the population). For most crocodilians, including the Chinese alligator, this is considered to be approximately 15 years. Therefore, a period of 3 generations is 45 years, which, using 2005 as a reference point, would extend our period of analysis back to 1960. Because there are no alligator census data from this period, we must base our analysis on inferred or suspected population size in 1960. We estimate that from 1979 to 2003 there was a population decline from approximately 1,000 to 100 alligators, a 90% decline over a 24–year period. If this estimated annual rate of decline (−3.75%) were used for the 19–year period from 1960 to 1979, the number of alligators in 1960 would have been approximately 1,713. But if we use 2008 as our reference point and propose that the Chinese alligator would no longer qualify as CR 15 years from now, that would be in 2023, and our reference year for 3 generations would then be 1978. The estimated population size in that year would be 1,038 alligators. Thus in order to have a less than 80% population decline over a period of 3 generations by 2023, the total number of wild Chinese alligators would need to be at least 207 individuals. This benchmark should be easily attainable.

To meet the conditions of criterion B for removal from CR, Chinese alligators would require an area of occupancy greater than 10 km² and have wild populations in at least two locations. To meet criteria C and D the wild population would have to exceed 250 adults, with at least one population having more than 50 adults. Criterion E is based on quantitative modeling and would require that the wild population have less than a 50% chance of extinction in the wild over the next 3 generations (45 years). One other extremely important point is that because recovery plans are based largely on releasing captive-reared animals, the IUCN (2001), stipulates that reintroduced individuals must have produced viable offspring before they are counted as mature individuals.

Summarizing the most critical points, Chinese alligators would no longer qualify for a CR listing if the number of wild adults exceeded 250, with at least one population having in excess of 50 adults, and the area of occupancy was greater than 10 km² (1,000 ha). In order to meet the criterion that 250 adults have successfully reproduced, it must be confirmed that at least 125 different females have nested. Furthermore, in order for the Chinese alligator to be downlisted to endangered by 2023, these criteria will have to be met by 2018 (as they have to be in place for 5 years prior to downlisting). Can these criteria be met? Given that the current size of the breeding groups in the NCAR does not exceed 2 adult females at any one site, it would be difficult to justify individual sites as "locations" under the IUCN criteria. We consider it to be more reasonable, at present, to consider the NCAR as a single location. Meeting the IUCN criteria for downlisting would then require the establishment of populations of alligators at one or more other sites. At present the only active program is at in Chongming Island, Shanghai. (In our opinion, the plans for the Changxing reintroduction do not meet the criteria for classification as a wild population.) The total size of the area of occupancy of alligators in the NCAR will be less than 300

The IUCN Red List Categories

Extinct (EX)

A taxon is "extinct" when there is no reasonable doubt that the last individual has died. A taxon is presumed extinct when exhaustive surveys in known and expected habitat, at appropriate times (diurnal, seasonal, annual), throughout its historic range have failed to record an individual. Surveys should be over a time frame appropriate to the taxon's life cycles and life form.

Extinct in the wild (EW)

A taxon is "extinct in the wild" when it is known to survive only in cultivation, in captivity, or as a naturalized population (or populations) well outside its past range. A taxon is presumed extinct in the wild when exhaustive surveys in known or expected habitat, at appropriate times (diurnal, seasonal, annual), throughout its historic range have failed to record an individual. Surveys should be over a time frame appropriate to the taxon's life cycle and life form.

Critically endangered (CR)

A taxon is "critically endangered" when the best available evidence indicates that it meets any of the criteria (A–E) for critically endangered and is therefore considered to be facing an extremely high risk of extinction in the wild.

Endangered (EN)

A taxon is "endangered" when the best available evidence indicates that it meets any of the criteria (A–E) for endangered and is therefore considered to be facing a very high risk of extinction in the wild.

Vulnerable (VU)

A taxon is "vulnerable" when the best available evidence indicates that it meets any of the criteria (A–E) for vulnerable and is therefore considered to be facing a high risk of extinction in the wild.

Near Threatened (NT)

A taxon is "near threatened" when it has been evaluated against the criteria and does not qualify for critically endangered, endangered, or vulnerable, but is close to qualifying for or is likely to qualify for a threatened category in the near future.

Least Concern (LC)

A taxon is "least concern" when it has been evaluated against the criteria and does not qualify for critically endangered, endangered, vulnerable, or near threatened. Widespread and abundant taxa are included in this category.

Data Deficient (DD)

A taxon is "data deficient" when there is inadequate information to make a direct or indirect assessment of its risk of extinction based on its distribution, population status, or both. A taxon in this category may be well studied and its biology well known, but appropriate data on abundance and distribution are lacking. Data deficient is not a category of threat. Listing taxa in this category indicates that more information is required and acknowledges the possibility that future research will show that a threatened classification is appropriate. It is important to make positive use of whatever data are available. In many cases great care should be exercised in choosing between DD and a threatened status. If the range of a taxon is suspected to be relatively circumscribed, if a considerable period of time has elapsed since the last record of the taxon, a threatened status may well be justified.

Not Evaluated (NE)

A taxon is "not evaluated" when it is has not yet been evaluated against the criteria.

The five criteria are

A. declining population (past, present, and/or projected)
B. geographic range size, and fragmentation, decline, or fluctuations
C. small population size, and fragmentation, decline, or fluctuations
D. very small population or very restricted distribution
E. quantitative analysis of extinction risk (e.g., population viability analysis)

ha, with Gaojinmiao (ca. 230 ha) by far the largest site. Preliminary plans call for a target population size of about 100 adults at Gaojinmiao, but it is unlikely that there would be more than 10 nesting females (= 20 reproductive adults) by 2018. Other sites in the NCAR could be expected to have a similar number of adult alligators. The Chongming site, if it succeeds in establishing a breeding alligator population, will be at least 120 ha, and preliminary plans call for a target population of 80 adult animals. By 2018 it may be possible to have an adult female breeding population of 10 (= 20 reproductive adults).

The Chongming wetland park is planned to reach 860 ha, and while not all of this would be suitable for alligator habitat, if only half was wetland this site could conceivably support a much larger population of adult alligators. Unfortunately this is not likely to happen by 2018.

It seems feasible that by 2018 it would be possible to have 60 breeding adults between Gaojinmiao, other sites in the NCAR, and the Chongming Reserve. These areas would have a total area of occupancy of approximately 420 ha. To reach the goal of downlisting to endangered would require one or more additional sites of more than 580 ha in size where an additional 190 breeding adults can be established by 2018. If the SFA and provincial forestry departments accept the challenge of down-listing the Chinese alligator as we have proposed, the location of additional areas is clearly the main obstacle. But in our judgment it is one that can be easily overcome by including just one more site for alligator reintroductions — Xiazhu Hu. The core area of the Xiazhu Hu wetland covers 1,150 ha, by far the largest site that we have currently identified and one that could support an estimated 460 adult Chinese alligators. Given the nature of the habitat at Xiazhu Hu, with large numbers of small islands that could serve as nesting habitat, this site could conceivably support an even larger population than predicted based on our estimate of 2.5 ha per adult. The total size of the Xiazhu Hu wetland (3,600 ha) is more than three times the size of the core area, and if alligators were also established in the reserve buffer zones, at lower densities, the entire adult population in Xiazhu Hu could be in excess of 1,000. If large-scale reintroductions of adult alligators were started by 2010, it is quite possible that nesting levels would approach the necessary value of 95 (= 190 breeding adults) within an eight-year period.

Establishing a Viable Wild Population of Chinese Alligators

Broadly defined, a viable population is one that maintains its vigor and potential for evolutionary adaptation (Soulé 1987). There is no single number that delimits a "viable" population, but rather a range of values that depends on how the minimum viable population is defined. Examples include a population having a 100% probability of surviving 1,000 years, a 90% chance of surviving for 100 years, or a 95% probability of surviving for 50 generations. Estimates of an MVP depend on demographic, genetic, and environmental factors and the kinds and levels of stochasticity that are present. The process of evaluating the long-term probability of survival of a species, referred to as a population viability analysis, usually requires fairly detailed demographic information on the species in question and a series of educated guesses about the severity and frequency of natural disasters such as floods and droughts. Many of the biological and environmental factors needed to conduct a PVA for Chinese alligators are poorly understood or unknown. A preliminary PVA carried out on Chinese alligators has emphasized their extreme vulnerability to catastrophic events and the urgent need to establish additional wild populations (Wu and Wang, pers. comm.).

Conducting a more detailed PVA of a species for which relatively little information is available can lead to unreliable predictions (Beissinger and Westphal 1998). Under these circumstances the most useful methods for evaluating population viability may be with general "rules of thumb" (Shaffer et al. 2002) or techniques based on expert opinion and habitat analysis (Groves 2003). There has been reluctance to use of rules of thumb or generalized estimates of MVP for a broad range of species because it has been assumed that there will be significant interspecific differences. However, an analysis by Reed et al. (2003) found that this might not necessarily be true and that existing MVP estimates did not vary significantly among major taxa. Early rule of thumb estimates for minimum viable population size concluded that over the short term the effective population size should be at least 50, and over the long term a minimum of 500 (Franklin 1980; Soulé 1980). Subsequent determinations have tended to increase these values. An effective population of approximately 1,000 (corresponding to more than 5,000 individuals) has been suggested as a minimum value to allow long-term adaptive evolution and the buildup of harmful mutations (Allendorf and Ryman 2002). PVA analyses of 102 vertebrate species by Reed et al. (2003) produced somewhat similar mean MVP of 7,316. While these latter authors do not claim to have come up with a magic number that will ensure long-term persistence of populations, they suggest a value of approximately 7,000 breeding-age adults as a general rule of thumb for MVP for vertebrates, and this is what we use as a final goal for Chinese alligator conservation efforts. Based on a rough estimate of 30% of the total (nonhatchling) crocodilian population being comprised of adults, this would correspond to an overall population of about 20,000 alligators. Assuming a 50:50 sex ratio and an average of one half of the adult females nesting in any one year, the target would be for an annual production of 1,750 nests.

Because of the alligator's extremely adaptable nature and its relatively modest area requirements, we feel that given proper support from the Chinese government this goal is clearly attainable. Our rough guidelines for the amount of land required for an adult pair of alligators is 2.5 ha. Assuming an even sex ratio, this suggests that a minimum of 8,750 ha of suitable habitat is needed for a target population of 7,000 adult alligators.

PROGNOSIS

The status of the Chinese alligator is an amalgam of contradictions. It is among the last surviving remnants of the formerly spectacular megafauna once spread across the lower Yangtze valley, but it teeters on the brink of extinction in the wild. There is a large national alligator reserve, but natural habitat is nonexistent. The wild population is tiny and highly fragmented, but a large number of captive animals exist in China and to a lesser degree in zoos outside the country. The Chinese government has considered the Chinese alligator to be a success story, but the wild population is critically endangered.

A numerical description of the status of the wild alligator population is disturbing. Most wild alligators exist as single individuals lost in an agricultural landscape. Natural reproduction appears to be limited to fewer than 10 sites at present, with no more than two nests at any site annually. Areas where alligators survive are highly altered and either marginal or unsuitable in terms of habitat quality and quantity. Regardless of the hatching success of these nests, the survival of hatchling alligators is questionable, and efforts to increase the survivorship of hatchlings will only provide limited benefits due to the physical limits in size of the existing sites. The remaining wild alligators are now in a downward spiral, succumbing to human population pressures: unintentionally poisoned, accidentally shot by hunters or drowned by fishermen. When combined with the absence of natural habitat, the species may be considered already functionally extinct in the wild. This became frighteningly clear to us in 1998 when there was not one nest laid in the wild.

The Chinese alligator offers us a vision of the future of conservation for many species as habitats dwindle and more emphasis is placed on ex situ programs for wildlife. For some species, the distinction between what is a wild animal and what is a captive individual will blur, and one of the most important lessons from out work with the Chinese alligator is the need to maintain a clear vision of this distinction. China has placed considerable emphasis on captive breeding for the conservation of the Chinese alligators, as well as a host of other species. Throughout this book, we stress the need to develop a conservation strategy that makes use of captive breeding to ensure the survival of the Chinese alligator *and* as a foundation upon which to build efforts to restore wild populations.

The future of Chinese alligators in the wild now depends on a two-part process: identifying and rehabilitating areas of adequate habitat, and establishing new wild alligator populations at these sites by releasing captive-bred alligators. Here the alligator gives cause for optimism. It is a highly adaptable species that does well in captivity, as evidenced by the success of the Chinese breeding program, as well as in a variety of wetland habitats. Preliminary results of trial reintroduction programs at three different sites have all shown that provided certain minimum standards are maintained, reintroductions can be an effective way to restore populations outside the breeding centers. The main limitation is habitat, but even in densely settled Eastern China there are options in remnant or restored wetlands in parts of the species' historic range. While some species, including most larger mammals, need reserves on the order of thousands of hectares to establish breeding populations, alligators can get by with much less space. Perhaps the best remaining surviving group of alligators (now producing two nests a year), is located in a pond whose area totals 8 ha. Much of the future success of the reintroduction program will rest on the development of new and novel approaches to the conservation and management of wetlands in Eastern China, and how well the establishment of new alligator populations can be linked with benefits to local communities. Wetlands conservation programs are set to grow in the coming years as the connection between wetlands and flood control and freshwater supplies is accentuated. Because, for the foreseeable future, many of the sites where alligators can be reestablished will be small (< 100 ha), alligators will likely exist as a network of small subpopulations that will require coherent management, genetically and demographically, among wild animals and the breeding centers. Leadership for this work is needed on many levels, but particularly from the national and provincial governments in China.

Literature Cited

Abercrombie, C. L., K. G. Rice, and C. A. Hope. 2001. The great alligator-caiman debate: Meditations on crocodilian life-history strategies. In *Crocodilian Biology and Evolution*, ed. G. C. Grigg, F. Seebacher, and C. E. Franklin, 409–418. Surrey Beatty and Sons, Chipping Norton, Australia.

Allendorf, F. W., and G. Luikart. 2007. *Conserving Global Biodiversity? Conservation and Genetics of Populations*. Blackwell Publishing, Oxford, UK. 642 pp.

Allendorf, F. W. and N. Ryman. 2002. The role of genetics in population viability analysis. In *Population Viability Analysis*, ed. S. R. Beissinger and D. R. McCullough, 50–85. University of Chicago Press, Chicago.

Anderson, J. 1898. *Zoology of Egypt: Reptilia and Bactracia*. Bernard Quaritch, London. 371 pp.

Anonymous. 1889. Chinese alligators in the aquarium at Berlin. *Scientific American* 60 (18): 278.

Bakken, D. A. 1997. Taphonomic parameters of Pleistocene hominid sites in China. *Indo-Pacific Prehistory Association Bulletin* 16: 13–26.

Barbour, T. 1910. A note regarding the Chinese alligator. *Proceedings of the Academy of Natural Sciences Philadelphia* 62: 464–467.

———. 1922. Further remarks on the Chinese alligator. *Proceedings of the New England Zoology Club* 8: 31–34.

Barton, A. J. 1955. Prolonged survival of a released alligator in Pennsylvania. *Herpetologica* 11: 210.

Bartram, W. 1791. *Travels through North and South Carolina....* James and Johnson, Philadelphia. 534 pp.

Bates, R. 2002. *Chinese Dragons*. Oxford University Press, Hong Kong. 96 pp.

Beck, B. B., L. G. Rapaport, M. R. Stanley Price, and A. C. Wilson. 1994. Reintroduction of captive-born animals. In *Creative Conservation*, ed. P. J. S. Olney, G. M. Mace, and A. T. C. Feistner, 265–286. Chapman and Hall, New York.

Behler, J. 1977. A propagation program for Chinese alligators (*Alligator sinensis*) in captivity. *Herpetology Review* 84: 124–125.

———. 1993. Chinese alligator (*Alligator sinensis*). AAZPA Annual Report on Conservation and Science, 1992–1993: 227–229.

———. 1995. Chinese alligator (*Alligator sinensis*). In AZA Annual Report on Conservation and Science, ed. K. Swaringen, R. J. Wiese, K. Willis, and M. Hutchins, 255–256. American Zoo and Aquarium Association, Bethesda, Md.

Beissinger, S. R., and M. I. Westphal. 1998. On the use of demographic models of population viability in endangered species management. *Journal of Wildlife Management* 62: 821–841.

Bellwood, P. 2005. Examining the farming/language dispersal hypothesis in the East Asian context. In *The Peopling of East Asia: Putting Together Archaeology, Linguistics, and Genetics*, ed. L. Sagart, R. Blench, and A. Sanchez-Mazas, 17–30. Routledge Curzon, New York.

Bennett, A. F., and K. A. Nagy. 1977. Energy expenditure of free-ranging lizards. *Ecology* 58: 697–700.

Birkhead, W. S., and C. R. Bennett. 1981. Observations of a small population of estuarine-inhabiting alligators near Southport, North Carolina. *Brimleyana* 6: 111–117.

Bloss, L. W. 1973. The Buddha and the nāga: A study in Buddhist folk religiosity. *History of Religions* 13: 36–53.

Boulenger, G. A. 1890. Remarks on the Chinese alligator. *Proceedings of the. Zoological Society of London* 1890: 619–620.

Brandt, L. A., and F. J. Mazzotti. 1990. The behavior of juvenile *Alligator mississippiensis* and *Caiman crocodilus* exposed to low temperature. *Copeia* 1990: 867–871.

Brazaitis, P. 1968. T'o. *Animal Kingdom* 1968: 24–27.

Bretschneider, E. 1898. *History of European Botanical Discoveries in China.* Sampson Low, Marston, London. 1184 pp.

Brisbin, I. L., Jr. 1990. Growth curve analyses and their application to the conservation and management of crocodilians. In *Crocodiles. Proceedings of the 9th Working Meeting of the Crocodile Specialist Group*, 116–145. IUCN–World Conservation Union, Gland, Switzerland.

Brisbin, I. L., Jr., E. A. Standora, and M. J. Vargo. 1982. Body temperature and behavior of American alligators during cold winter weather. *American Midland Naturalist* 107: 209–218.

Britton, A. R. C. 2001. Review and classification of call types of juvenile crocodilians, and factors affecting distress calls. In *Crocodilian Biology and Evolution*, ed. G. C. Grigg, F. Seebacher, and C. E. Franklin, 364–377. Surrey Beatty and Sons, Chipping Norton, Australia.

Brochu, C. A. 1999. Phylogenetics, taxonomy, and historical biogeography of Alligatoroidea. Society of Vertebrate Paleontology Memoir 6. *Journal of Vertebrate Paleontology*, supplement to no. 2, 19: 9–100.

———. 2000. Phylogenetic relationships and divergence timing of *Crocodylus* based on morphology and the fossil record. *Copeia* 2000: 657–673.

———. 2003. Phylogenetic approaches toward crocodylian history. *Annual Review of Earth and Planetary Science* 31: 357–397.

Burke, R. L. 1991. Relocations, repatriations, and translocations of amphibians and reptiles: Taking a broader view. *Herpetologica* 47: 350–357.

Campbell, H. W. 1973. Observations on the acoustic behavior of crocodilians. *Zoologica* 58: 1–11.

Campbell, M. R., and F. J. Mazzotti. 2004. Characterization of natural and artificial alligator holes. *Southeastern Naturalist* 3: 583–594.

Carr, A. 1963. *The Reptiles.* Time, New York. 192 pp.

Caughley, G. 1994. Directions in conservation biology. *Journal of Animal Ecology* 63: 215–244.

Chabreck, R. H. 1966. Methods of determining the size and composition of alligator populations in Louisiana. *Proceedings Annual Conference Southeastern Association Game and Fish Commissions.* 20: 105–112.

———. 1975. Moisture variation in nests of the American alligator, *Alligator mississippiensis. Herpetologica* 31: 385–389.

Chabreck, R. H., and T. Joanen. 1979. Growth rates of American alligators in Louisiana. *Herpetologica* 35: 51–57.

Chambers, R. 1881. *The Book of Days.* W and R Chambers, Edinburg. 843 pp.

Chang, K.-C. 1987. *The Archaeology of Ancient China.* Yale University Press, New Haven. 483 pp.

Chang, K.-C., and P. Xu (eds.) 2005. *The Formation of Chinese Civilization: An Archaeological Perspective.* Yale University Press, New Haven. 363 pp.

Chang, T. T. 1988. Ethnobotany of rice in insular Southeast Asia. *Asian Perspectives* 26: 69–76.

Chen, B. 1985. *Chinese Alligator.* Anhui Science and Technology Press, Hefei, Anhui, PRC. 519 pp.

———. 1990a. The past and present situation of the Chinese alligator. *Asiatic Herpetological Research* 3: 129–136.

———. 1990b. TDSD in Chinese alligators. *Crocodile Specialist Group Newsletter* 9: 11.

———. 1991. Chinese alligator. In *The Amphibian and Reptilian Fauna of Anhui*, ed. B. Chen, 361–365. Anhui Publishing House of Science and Technology. Hefei, Anhui, China.

———. 1998. *Alligator sinensis* Fauvel, 1879. In *China Red Data Book of Endangered Animals: Amphibia and Reptilia*, ed. E. Zhao, 311–319. Science Press, Beijing.

Chen, B., T. Hua, X. Wu, and C. Wang 2003. *Research on the Chinese Alligator.* Shanghai Scientific and Technological Education Publishing House. Shanghai, China. 519 pp.

Chen, B., and B. Li. 1979. Initial observations on ecology of the Chinese alligator. *Journal of Anhui Teacher's College*. 1: 69–73.

Chen, B., and C. Wang. 1984. Artificial reproduction of *Alligator sinensis*. *Acta Herpetologica Sinica* 3: 49–53.

Chen, J., and K. Chen. 1997. The conservation status of wetlands in China and their future prospects. In *Development Policies, Plans, and Wetlands: Proceedings of Workshop 1 of the International Conference on Wetlands and Development*, ed. R. C. Prentice and R. P. Jaensch, 117–126. Wetlands International, Kuala Lumpur.

Chen, K., and G. Zhang (eds.). 1998. Wetland and waterbird conservation. *Proceedings of an International Workshop on Wetland and Waterbird Conservation in North East Asia*. Wetland International: China Programme. China Forestry Publishing House, Beijing. China. 294 pp.

Chen, X., Y. Zong, E. Zhang, J. Xu, and S. Li. 2001. Human impacts on the Changjiang (Yangtze) River basin, China, with special reference to the impacts on the dry season water discharges into the sea. *Geomorphology* 41: 111–123.

Child, G. 1987. The management of crocodiles in Zimbabwe. In *Wildlife Management: Crocodiles and Alligators*, ed. G. J. W. Webb, S. C. Manolis, and P. Whitehead, 49–62. Surrey Beatty and Sons, Chipping Norton, Australia.

Chow, M., and B. Wang. 1964. Fossil vertebrates from the Miocene of Northern Kiangsu. *Vertebrata PalAsiatica* 8: 341–351.

Chu, C. 1957. Observations on the life history of the Chinese alligator. *Acta Zoologica Sinica* 92: 129–143.

Cintra, R. 1988. Nesting ecology of the Paraguayan caiman (*Caiman crocodilus yacare*) in the Brazilian Pantanal. *Journal of Herpetology* 22: 219–222.

Clermont-Ganneau, C. 1877. *Horus et Saint Georges*. Paris: Librairie Académique Didier.

Cohen, A. 1978. Coercing the rain deities in ancient China. *History of Religions* 17: 244–265.

Conner, R. N. 1988. Wildlife populations: Minimally viable or ecologically functional? *Wildlife Society Bulletin* 16: 80–84.

Cott, H. B. 1961. Scientific results of an inquiry into the ecology and economic status of the Nile crocodile (*Crocodilus niloticus*) in Uganda and Northern Rhodesia. *Transactions of the Zoological Society of London* 29: 211–358.

Craig, J. L. 1994. Meta-populations: Is management as flexible as nature? In *Creative Conservation: Interactive Management of Wild and Captive Animals*, ed. P. J. S. Olney, G. M. Mace, and A. T. C. Feistner, 50–66. Chapman and Hall, London.

Craighead, F. C. 1968. The role of the alligator in shaping plant communities and maintaining wildlife in the southern Everglades. *Florida Naturalist* 41: 2–7.

Crawshaw, P. G. J. 1987. Nesting ecology of the Paraguayan caiman (*Caiman yacare*) in the Pantanal of Mato Grosso, Brazil. Master's thesis. University of Florida, Gainesville. 69 pp.

Crocodile Specialist Group.1983. China. *Crocodile Specialist Group Newsletter* 2: 5.

———. 1984. China. *Crocodile Specialist Group Newsletter* 3: 5.

Dalrymple, G. H. 1996. Growth of American alligators in the Shark Valley region of Everglades National Park. *Copeia* 1996: 212–216.

Da Silveira, R., and J. B. Thorbjarnarson. 1999. Conservation implications of commercial hunting of black and spectacled caiman in the Mamirauá Sustainable Development Reserve, Brazil. *Biological Conservation* 88: 103–109.

Davenport, M. 1982. Natural history notes: *Alligator sinensis*, reproduction/aging. *Herpetological Review* 13: 94–95.

Davis, L. M. 2002. Genetic variation, mating patterns, and reproductive dynamics in American alligators. PhD diss., University of South Carolina, Columbia. 256 pp.

Davis, L. M., T. C. Glenn, R. M. Elsey, H. C. Dessauer, and R. H. Sawyer. 2001. Multiple paternity and mating patterns in the American alligator, *Alligator mississippiensis*. *Molecular Ecology* 10: 1011–1024.

de Boer, L. E. M. 1994. Development of coordinated genetic and demographic breeding programmes. In *Creative Conservation: Interactive Management of Wild and Captive Animals*, ed. P. J. S. Olney, G. M. Mace, and A. T. C. Feistner, 304–311. Chapman and Hall, London. 517 pp.

Deitz, D. C. 1979. Behavioral ecology of young American alligators. PhD diss., University of Florida, Gainesville. 161 pp.

Deitz, D. C., and T. C. Hines. 1980. Alligator nesting in north-central Florida. *Copeia* 1980: 249–258.

Delany, M. F., S. B. Linda, and C. T. Moore. 1999. Diet and condition of alligators in 4 Florida lakes. *Proceedings of the Annual Conference Southeastern Association of Fish and Wildlife Agencies* 53: 375–389.

de Ricqles, A. J. 1975. On bone histology of fossil and living reptiles, with comments on its functional and evolutionary significance. In *Morphology and Biology of Reptiles*, ed. A. Bellairs and C. B. Cox, 211–216. Academic Press, New York.

de Visser, M. W. 1913. *The Dragon in China and Japan*. J. Müller, Amsterdam. 243 pp.

Ding, X. 2001. General planning of the reintroduction of Chinese alligator in Yixing County, Jiangsu Province. In *Status Quo and Future of Conservation for Chinese Alligator and Crocodiles in the World: Proceedings of the International Workshop on Conservation and Reintroduction of Chinese Alligator, Hefei, China, 2001, and International Workshop on Captive Breeding and Commerce Management in Crocodylia, Guangzhou, China, 2001*. ed. W. Wang, X. Ruan, P. Si, D. Zhang, Y. Luo, C. Wang, Q. Sun, F. Qian, and H. Jiang. 180–192. China Forestry Publishing House, Beijing.

Ding, Y., X. Wang, L. He, Z. Wang, W. Wu, F. Tao, and M. Shao. 2003. Position of burrow entrances in wild Chinese alligators. *Zoological Research* 24: 254–258. (In Chinese with English abstract.)

Ding, Y., X. Wang, J. Thorbjarnarson, J. Wu, Z. Wang, W. Wu, C. Gu, and J. Nie. 2004. Movement patterns of released captive-reared Chinese alligators (*Alligator sinensis*). In *Crocodiles. Proceedings of the 17th Working Meeting of the Crocodile Specialist Group*, 109. IUCN–World Conservation Union, Gland, Switzerland.

Dodd, C. K., and R. A. Seigel. 1991. Reintroductions, translocations, and repatriations: Are they effective conservation strategies? *Herpetologica* 47: 336–350.

Durand, J. D. 1960. The population statistics of China, A.D. 2–1953. *Population Studies* 13: 209–256.

Dutton, H. J., A. M. Brunell, D. A. Carbonneau, L. J. Hord, S. G. Stiegler, C. H. Visscher, J. H. White, and A. R. Woodward. 2002. Florida's alligator management program: An update 1987 to 2001. In *Crocodiles: Proceedings of the Sixteenth Working Meeting of the Crocodile Specialist Group*, 23–30. IUCN–World Conservation Union, Cambridge, Eng.

Ebrey, P. B. 1996. *Cambridge Illustrated History: China*. Cambridge University Press, Cambridge, Eng. 352 pp.

Elsey, R. M. 2005. Unusual offshore occurrence of an American alligator. *Southeastern Naturalist* 4: 533–536.

Elsey, R. M., and N. Kindler. 2007. The American alligator: A sustainable use success story. *Species* 48: 11–14.

Elvin, M. 2004. *The Retreat of the Elephants: An Environmental History of China*. Yale University Press, New Haven. 592 pp.

Ewert, M., and C. Nelson. 2003. Metabolic heating of embryos and sex determination in the American alligator, *Alligator mississippiensis*. *Journal of Thermal Biology* 28: 159–165.

Fan, F. 2004. *British Naturalists in Qing China*. Harvard University Press, Cambridge, Mass. 253 pp.

Fang, J., Z. Wang, S. Zhao, Y. Li, Z. Tang, D. Yu, L. Ni, H. Liu, P. Xie, L. Da, Z. Li, and C. Zheng. 2006. Biodiversity changes in the lakes of the central Yangtze. *Frontiers in Ecology and the Environment* 4: 369–377.

Fang, S. 2001. Action plan for returning *Alligator sinensis* to the field in Changxing County, Zhejiang Province. In *Status Quo and Future of Conservation for Chinese Alligator and Crocodiles in the World: Proceedings of the International Workshop on Conservation and Reintroduction of Chinese Alligator, Hefei, China, 2001, and International Workshop on Captive Breeding and Commerce Management in Crocodylia, Guangzhou, China, 2001,* ed. W. Wang, X. Ruan, P. Si, D. Zhang, Y. Luo, C. Wang, Q. Sun, F. Qian, and H. Jiang, 164–171. China Forestry Publishing House, Beijing.

Fang, S., X. Xu, X. Yu and Z. Li. 2005. Poplar in wetlands agroforestry: A case study of ecological benefits, site productivity, and economics. *Wetlands Ecology and Management* 13: 93–104.

Fauvel, A. A. 1879. Alligators in China. Privately printed monograph. 36 pp. Published in 1879 as Alligators in China in *Journal of the North-China Branch of the Royal Asiatic Society, Shanghai* 13.

Ferguson, M. W. J. 1985. Reproductive biology and embryology of the crocodilians. In *Biology of the Reptilia.* Vol. 14, *Development A.* ed. C. Gans, F. S. Billet, and P. F. A. Maderson, 330–491. John Wiley and Sons, New York.

Ferguson, M. W. J., and T. Joanen. 1983. Temperature-dependent sex determination in *Alligator mississippiensis. Journal of Zoology* 200: 143–177.

Fischer, J., and D. B. Lindenmayer. 2000. An assessment of the published results of animal relocations. *Biological Conservation* 96: 1–11.

Fitzgerald, L. A. 1988. Dietary patterns of *Caiman crocodilus* in the Venezuelan llanos. Master's thesis, University of New México, Albuquerque. 74 pp.

Foot, C. 1996. Crocodile exports to China. *Crocodile Specialist Group Newsletter* 15: 5–6.

Frankham, R. 1995. Effective population size/adult population size ratios in wildlife: A review. *Genetical Research* 66: 95–107.

Frankham, R., J. D. Ballou, and D. A. Briscoe. 2002. *Introduction to Conservation Genetics.* Cambridge University Press, Cambridge, Eng.

Franklin, I. R. 1980. Evolutionary change in small populations. In *Conservation Biology: An Evolutionary-Ecological Perspective,* ed. M. E. Soulé and B. A. Wilcox, 135–150. Sinauer, Sunderland, Mass.

Frederick, P., and M. Spaulding. 1994. Factors affecting reproductive success of wading birds (Ciconiiformes) in the Everglades ecosystem. In *Everglades: The Ecosystem and Its Restoration,* ed. S. M. Davis and J. C. Ogden, 659–691. St. Lucie Press, Delray Beach, Fla.

Fu, C., J. Wu, J. Chen, Q. Wu and G. Lei. 2003. Freshwater fish biodiversity in the Yangtze River basin of China: Patterns, threats, and conservation. *Biodiversity and Conservation* 12: 1649–1685.

Fu, J. 1994. Conservation, management, and farming of crocodiles in China. In *Crocodiles: Proceedings of the 2nd Regional (Eastern Asia, Oceania, Australasia) Meeting of the Crocodile Specialist Group.* IUCN–World Conservation Union, Gland, Switzerland.

Fuller, M. K. 1981. Characteristics of an American alligator (*Alligator mississippiensis*) population in the vicinity of Lake Ellis Simon, North Carolina. Master's thesis, North Carolina State University, Raleigh. 136 pp.

Garrick, L. D. 1975. Structure and pattern of the roars of Chinese alligators (*Alligator sinensis* Fauvel). *Herpetologica* 31: 26–31.

Garrick, L. D., and J. Lang. 1975. Alligator courtship. *American Zoologist* 15: 813.

———. 1977. Social signals and behaviors of adult American alligators. *American Zoologist* 17: 225–239.

Garrick, L. D., J. Lang, and H. A. Herzog. 1978. Social signals of adult American alligators. *Bulletin of the American Museum of Natural History* 160: 157–192.

Gatesy, J., G. Amato, M. Norell, R. DeSalle, and C. Hayashi. 2003. Combined support for wholesale taxic atavism in gavialine crocodylians. *Systematic Biology* 52: 403–422.

Gharial Working Group. 2007. Indian gharial listed as "Critically Endangered" in IUCN Red List. *Crocodile Specialist Group Newsletter* 26: 7–9.

Gholz, H. L., C. S. Perry, W. P. Cropper Jr., and C. Hendry. 1985. Litterfall, decomposition, and nitrogen and phosphorus dynamics in a chronosequence of slash pine (*Pinus elliottii*) plantations. *Forest Science* 31: 463–478.

Gilligan, D. M., and R. Frankham. 2003. Dynamics of genetic adaptation to captivity. *Conservation Genetics* 4: 189–197.

Gilmore, D. D. 2003. *Monsters: Evil Beings, Mythical Beasts, and All Manner of Imaginary Terrors.* University of Pennsylvania Press, Philadelphia. 232 pp.

Glasgow, V. L. 1991. *A Social History of the American Alligator.* St. Martin's Press, New York. 260 pp.

Glenn, T. C., J. L. Staton, A. T. Vu, L. M. Davis, J. R. A. Bremer, W. E. Rhodes, I. L. Brisbin, and R. H. Sawyer. 2002. Low mitochondrial DNA variation among American alligators and a novel non-coding region in crocodilians. *Journal of Experimental Zoology* 294: 312–324.

Goodwin, T., and W. R. Marion. 1979. Seasonal activity ranges and habitat preferences of adult alligators in a north-central Florida lake. *Journal of Herpetology* 13: 157–164.

Gordon, C. A. 1884. *An Epitome of the Reports of the Medical Officers to the Chinese Imperial Maritime Customs Service from 1871 to 1882.* Bailliere, Tindal, and Cox, London. 435 pp.

Gore, R. 1978. A bad time to be a crocodile. *National Geographic* 153: 90–115.

Grigg, G. C., and F. Seebacher. 2001. Crocodilian thermal relations. In *Crocodilian Biology and Evolution*, ed. G. C. Grigg, F. Seebacher, and C. E. Franklin, 297–309. Surrey Beatty and Sons, Chipping Norton, Australia.

Groombridge, B. 1982. *Amphibia-Reptilia Red Data Book. Part 1, Testudines, Crocodilia, and Rhynchocephalia.* IUCN–World Conservation Union, Gland, Switzerland. 426 pp.

Groves, C. R. 2003. *Drafting a Conservation Blueprint.* Island Press, Washington, D.C. 457 pp.

Gu, C., R. Wang, C. Wang, and Y. Zhou. 2000. Medicinal value and conservation of Chinese alligators. In *Conservation of Endangered Medicinal Wildlife Resources in China*, 114–121. Second Military Medical University Press, Shanghai. (In Chinese with English abstract.)

Guggisberg, C. A. W. 1972. *Crocodiles: Natural History, Folklore, and Conservation.* Stockpole Books, Harrisburg, Pa. 195 pp.

Guillette, L. J. J., A. R. Woodward, D. A. Crain, G. R. Masson, P. D. Palmer, M. C. Cox, X. Y. Qui, and E. F. Orlando. 1997. The reproductive cycle of the female American alligator (*Alligator mississippiensis*). *General and Comparative Endocrinology* 108: 87–101.

Gupta, R. C., and C. S. Bhardwaj. 1995. Investigation on the tunnel ecology of *Crocodylus palustris* Lesson. *Journal of Environmental Biology* 16: 167–174.

Hagan, J. M. 1982. Movement habits of the American alligator (*Alligator mississippiensis*) in North Carolina. Master's thesis, North Carolina State University, Raleigh. 203 pp.

Hagan, J. M., P. C. Smithson, and P. D. Doerr. 1983. Behavioral response of the American alligator to freezing weather. *Journal of Herpetology* 17: 402–404.

Hanks, L. M. 1992. *Rice and Man.* University of Hawai'i Press, Honolulu. 196 pp.

Hanski, I, and M. Gilpin. 2008. Metapopulation dynamics: Brief history and conceptual domain. *Biological Journal of the Linnean Society* 42: 3–16.

Harris, Richard. 2005. *Wildlife Conservation in China: Preserving the Habitat of China's Wild West.* M. E. Sharpe. New York. 341 pp.

Hartwell, R. M. 1982. Demographic, political, and social transformations of China, 750–1550. *Harvard Journal of Asiatic Studies* 42: 365–442.

He, X. 1986. *Origin of Gods: Chinese Ancient Mythology and History.* Beijing: Sanlian Bookstore. (In Chinese.)

———. 1989. *Tan Long (On Dragons).* Hong Kong, Zhonghua Shuju Book Co.

Herbert, J. D., T. D. Coulson, and R. A. Coulson. 2002. Growth rates of Chinese and American alligators. *Comparative Biochemistry and Physiology Part A: Molecular and Integrative Physiology* 131: 909–916.

Herzog, H. A. 1974. The vocal communication system and related behaviours of the American alligator (*A. mississipiensis*) and other crocodilians. Master's thesis, University of Tennessee, Knoxville. 83 pp.

Herzog, H. A., and G. M. Burghardt. 1977. Vocalization in juvenile crocodilians. *Zeitschrift fur Tierpsychologie* 44: 294–304.

Hines, T. C., and C. L. Abercrombie. 1987. The management of alligators in Florida, USA. In

Wildlife Management: Crocodiles and Alligators, ed. G. J. W. Webb, S. C. Manolis, and P. Whitehead, 43–47. Surrey Beatty and Sons, Chipping Norton, Australia.

Hines, T. C., M. J. Fogarty, and L. C. Chappell. 1968. Alligator research in Florida: A progress report. *Proceeding of the Southeastern Association of Game and Fish Commissions* 22: 166–180.

Hines, T. C., and F. H. Percival. 1987. Alligator management and value-added conservation in Florida, In *Valuing Wildlife Economic and Social Perspectives*, ed. D. J. Decker and G. R. Goff, 164–193. Westview Press, Boulder, Colo.

Hoffman, W., G. T. Bancroft, and R. J. Sawicki. 1994. Foraging habitat of wading birds in the Water Conservation Areas of the Everglades. In *Everglades: The Ecosystem and Its Restoration*, ed. S. M. Davis and J. C. Ogden, 585–614. St. Lucie Press, Delray Beach, Fla.

Hogarth, P. J. 1980. St. George: The evolution of a saint and his dragon. *History Today* 30: 17–22.

Hopkins, L. H. 1913. Dragon and alligator: Being notes on some ancient inscribed bone carvings. *Journal of the Royal Asiatic Society of Great Britain and Ireland* 1913: 545–552.

Hornblower, G. D. 1933. Early dragon forms. *Man* 33: 79–87.

Hsiao, S. D. 1934. Natural history notes on the Yangtze alligator. *Peking Natural History Bulletin* 9: 283–293.

Hsu, Y. 2002. *Ancient Chinese Writing: Oracle Bone Inscriptions from the Ruins of Yin*. Taipei, National Palace Museum. 133 pp.

Hua, T., C. Wang, and B. Chen. 2004. Stages of embryonic development for *Alligator sinensis*. *Zoological Research* 25: 263–271.

Huang, C. 1959. Supplement on Chinese alligator research. *Dogwu Zazhi* 3: 257–277.

———. 1982. The ecology of the Chinese alligator and changes in its geographical distribution. In *Crocodiles: Proceedings of the 5th Working Meeting of the IUCN/SSC Crocodile Specialist Group*, 54–62. IUCN–World Conservation Union, Gland, Switzerland.

———. 1983. China. *Crocodile Specialist Newsletter* 2: 5–6.

Huang, M., Y. Jin, and C. Cai. 1987. *Fauna of Zhejiang*. Zhejiang Science and Technology Publishing House, Hangzhou, Zhejiang.

Huang, W., F. Song, X. Guo, and D. Chen. 1988. First discovery of *Megalovis guangxiensis* and *Alligator* cf. *sinensis* in Guangdong. *Vertebrata PalAsiatica* 26: 227–231.

Huang, Z. 1981. The Chinese alligator. *Oryx*. 16: 139–140.

Huang, Z., H. Lin, and S. Zhang. 1985. An analysis of the remote sensing image of the Chinese alligator habitat. *Oceanologica et Limnologica Sinica* 16: 35–42.

Huang, Z., and M. E. Watanabe. 1986. Nest excavation and hatchling behaviors of Chinese alligator and American alligator. *Acta Herpetologica Sinica*. 5: 5–10.

Hunt, R. H., 1987. Nest excavation and neonate transport in wild *Alligator mississippiensis*. *Journal of Herpetology* 21: 348–350.

Hutton, J. M. 1986. Age determination of living Nile crocodiles from the cortical stratification of bone. *Copeia* 1986: 332–341.

———. 1987. Incubation temperatures, sex ratios, and sex determination in a population of Nile crocodiles (*Crocodylus niloticus*). *Journal of Zoology* (London) 211: 143–155.

Hutton, J. M., and G. Webb. 2002. Legal trade snaps back: Using the experience of crocodilians to draw lessons on regulation of the wildlife trade. In *Crocodiles: Proceedings of the 16th Working Meeting of the Crocodile Specialist Group*, 1–10. IUCN–World Conservation Union. Gland, Switzerland, and Cambridge, Eng.

Inskipp, T., and S. Wells. 1979. *International Wildlife Trade*. Earthscan, London.

Isbell, L. A. 2006. Snakes as agents of evolutionary change in primate brains. *Journal of Human Evolution* 31: 1–35

IUCN. 1998. Guidelines for re-introductions. Prepared for the IUCN/SSC Reintroduction Specialist Group, IUCN–World Conservation Union, Gland Switzerland, and Cambridge, Eng. 10 pp.

———. 2001. *IUCN Red List Categories and Criteria: Version 3.1*. IUCN Species Survival Commission. IUCN–World Conservation Union, Gland, Switzerland, and Cambridge, Eng.

Jacobsen, T., and J. A. Kushlan. 1989. Growth dynamics in the American alligator (*Alligator mississippiensis*). *Journal of Zoology* (London) 219: 309–328.

Janke, A., A. Gullberg, S. Hughes, R. K. Aggarwal, and U. Arnason. 2005. Mitogenomic analyses place the gharial (*Gavialis gangeticus*) on the crocodile tree and provide Pre-K/T divergence times for most crocodilians. *Journal of Molecular Evolution* 61: 620–626.

Janzen, F. J. 1995. Experimental evidence for the evolutionary significance of temperature sex determination. *Evolution* 49: 864–873.

Jelden, D. 1997. Live alligator exports authorized. *Crocodile Specialist Group Newsletter* 16: 12.

Jenkins, M., and S. Broad. 1994. *International Trade in Reptile Skins: A Review of the Main Consumer Markets, 1983–1991*. TRAFFIC International, Cambridge, UK. 68 pp.

Jiang, H. 2008. Reintroduction achievement for Chinese alligator. *Crocodile Specialist Group Newsletter* 27: 9–10.

Jiang, H., G. Chu, X. Ruan, X. Wu, K. Shi, J. Zhu, and Z. Wang. 2006. Implementation of China action plan for conservation and reintroduction of Chinese alligator. In *Crocodiles: Proceedings of the 18th Working Meeting of the Crocodile Specialist Group*, 322–332. IUCN–World Conservation Union, Gland, Switzerland, and Cambridge Eng.

Jiang, L., and L. Liu 2005 The discovery of an 8000-year-old dugout canoe at Kuahuqiao in the Lower Yangzi River, China. *Antiquity* 79, no. 305. http://www.antiquity.ac.uk/projgall/liu/index.html.

Jiang, Z., and H. Hu. 2001. Père David's deer in China. II. Field-released. www.chinabiodiversity.com/shengwudyx2/training/chapter14-2.htm.

Jiang, Z., C. Yu, Z. Feng, L. Zhang, J. Xia, Y. Ding and N. Lindsay. 2000. Reintroduction and recovery of Père David's deer in China. *Wildlife Society Bulletin* 28: 681–687.

Joanen, T., and L. McNease. 1972. A telemetric study of adult male alligators on Rockefeller Refuge, Louisiana. *Proceedings of the Annual Conference of Southeastern Association of Game and Fish Commissions* 26: 252–275.

———. 1980. Reproductive biology of the American alligator in southwest Louisiana. In *Reproductive Biology and Diseases of Captive Reptiles*, ed. J. B. Murphy and J. T. Collins, 153–160. Society for the Study of Reptiles and Amphibians, Lawrence, Kans.

———. 1981. Nesting chronology of the American alligator and factors affecting nesting in Louisiana. In *Proceedings of the First Annual Alligator Production Conference*, ed. P. Cardeilhac, T. Lane, and R. Larson, 107–116. University of Florida, Gainesville.

———. 1989. Ecology and physiology of nesting and early development of the American alligator. *American Zoologist* 29: 987–998.

Joanen, T., L. McNease, and J. Behler. 1980. Captive propagation of the Chinese alligator (*Alligator sinensis*). Paper presented at the Southern Zoo Workshop, Monroe, La. 1 p.

Joanen, T., L. McNease, R. M. Elsey, and M. A. Staton. 1997. The commercial consumptive use of the American alligator (*Alligator mississippiensis*) in Louisiana: Its effects on conservation. In *Harvesting Wild Species: Implications for Biodiversity Conservation*, ed. C. H. Freese, 465–506. Johns Hopkins University Press, Baltimore.

Joanen, T., L. McNease, and M. W. Ferguson. 1987. The effects of egg incubation temperature on post-hatching growth of American alligators. In *Wildlife Management: Crocodiles and Alligators*, ed. G. J. W. Webb, S. C. Manolis, and P. J. Whitehead, 533–537. Surrey Beatty, Chipping Norton, Australia.

Jones, David. 2000. *An Instinct for Dragons*. Routledge, New York. 203 pp.

Kahler, W. R. 1895. *My Holidays in China: Part of Three Houseboat Tours, from Shanghai to Hangchow and Back via Ninpo; from Shanghai to Le Yang via Soochow and the Tah Hu, and from Kiukiang to Wuhu*. Temperance Union, Shanghai. 180 pp.

Kellogg, R. 1929. The habits and economic importance of alligators. *Technical Bulletin, U.S. Department of Agriculture* 147: 1–36.

Khan, J., and J. Yardley. 2007 As China roars, pollution reaches deadly extremes. *New York Times*, August 26, 2007.

King, F. W. 1978. The wildlife trade. In *Wildlife and America*, ed. H. P. Brokaw, 253–161. Council on Environmental Quality, Washington, D.C.

Klause, S. E. 1983. *Reproductive characteristics of the American alligator (Alligator mississippiensis) in North Carolina*. Master's thesis, North Carolina State University, Raleigh. 85 pp.

Kleiman, D. G. 1989. Reintroduction of captive mammals for conservation. *BioScience* 39: 152–161.

Kleiman, D. G., and A. B. Rylands. 2002. *Lion Tamarins: Biology and Conservation*. Smithsonian Institution Press, Washington, D.C. 448 pp.

Klemens, M. W., and J. B. Thorbjarnarson. 1995. Reptiles as a food resource. *Biodiversity and Conservation* 4: 281–298.

Kofron, C. P. 1990. The reproductive cycle of the Nile crocodile (*Crocodylus niloticus*). *Journal of Zoology, London* 221: 477–488.

———. 1993. Behavior of Nile crocodiles in a seasonal river in Zimbabwe. *Copeia* 1993: 463–469.

Kushlan, J. A. 1973. Observations on maternal behavior in the American alligator, *Alligator mississippiensis*. *Herpetologica* 29: 256–257.

———. 1974. Observations on the role of the American alligator (*Alligator mississippiensis*) in the southern Florida wetlands. *Copeia* 1974: 993–996.

Laidler, L., and K. Laidler. 1996. *China's Threatened Wildlife*. Blandford, London. 192 pp.

Lance, V. A. 2003. Alligator physiology and life history: The importance of temperature. *Experimental Gerontology* 38: 801–805.

Lance, V. A., and M. H. Bogart. 1994. Studies on sex determination in the American alligator *Alligator mississippiensis*. *Journal of Experimental Zoology* 270: 79–85.

Lang, J. W. 1987. Crocodilian behavior: Implications for management. In *Wildlife Management: Crocodiles and Alligators*, ed. G. J. W. Webb, S. C. Manolis, and P. J. Whitehead, 273–294. Surrey Beatty and Sons, Chipping Norton, Australia.

Lang, J. W., and H. V. Andrews. 1994. Temperature-dependent sex determination in crocodilians. *Journal of Experimental Zoology* 270: 28–44.

Levins, R. 1968. *Evolution in Changing Environments: Some Theoretical Explorations*. Monograph in Population Biology, Princeton University Press, Princeton, N.J. 126 pp.

———. 1969. Some demographic and genetic consequences of environmental heterogeneity for biological control. *Bulletin of the Entomology Society of America* 71: 237–240.

Li, C., M. Shao, H. Zhu, and J. Nie. 1996. Current status of *Alligator sinensis*. *Chinese Biodiversity* 4: 83–86. (In Chinese.)

Li, J., and B. Wang. 1987. A new species of *Alligator* from Shanwang, Shandong. *Vertebrata PalAsiatica* 7: 199–207.

Li, P. J. 2007. Enforcing wildlife protection in China: The legislative and political solutions. *China Information* 21: 71–107.

Li, W., and X. Zhao. 1989. *China's Nature Reserves*. Foreign Language Press, Beijing. 191 pp.

Li, X., G. Harbottle, J. Zhang, and C. Wang. 2003. The earliest writing? Sign use in the seventh millennium BC at Jiahu, Henan Province, China. *Antiquity* 77: 31–44.

Li, Y., and D. S. Wilcove. 2005. Threats to vertebrate species in China and the United States. *Bioscience* 55: 147–153.

Lindburg, D., and K. Baragona (eds.) 2004. *Giant Pandas: Biology and Conservation*. University of California Press, Berkeley. 308 pp.

Lipsky, M. K., and M. F. Child. 2007. Combining the fields of reintroduction biology and restoration ecology. *Conservation Biology* 21: 1387–1388.

Liu, J., Z. Ouyang, S. Pimm, P. Raven, X. Wang, H. Miao and N. Han. 2003. Protecting China's biodiversity. *Science* 300: 1240–1241.

Liu, L. 2000. Ancestor worship: An archaeological investigation of ritual activities in Neolithic North China. *Journal of East Asian Archaeology*. 2: 129–164.

———. 2004. *The Chinese Neolithic*. Cambridge University Press, New York. 328 pp.

Liu, X., K. Wang, and G. Zhang. 2004. Perspectives and policies: Ecological industry substitutes in wetland restoration of the middle Yangtze. *Wetlands* 24: 633–641.

Lu, T. L. D., 1999. *The Transition from Foraging to Farming and the Origin of Agriculture in China.* British Archaeological Reports. BAR International Series 774. 233 pp.

————. 2005. The origin and dispersal of agriculture and human diaspora in East Asia. In *The Peopling of East Asia Putting Together Archaeology, Linguistics and Genetics,* ed. L. Sagart, R. Blench, and A. Sanchez-Mazas, pp 51–62. Routledge Curzon, New York.

Lu, W., X. Yu, and Y. Hua. 1988. Preliminary study on reproduction of *Alligator sinensis* in artificial growth condition. *Journal of Zhejiang Forestry Science and Technology* 8: 30–32.

Mackinnon, J., M. Sha, C. Cheung, G. Carey, X. Zhu, and D. Melville 1996. *A Biodiversity Review of China.* WWF-China, Hong Kong. 558 pp.

Magnusson, W. E. 1986. The peculiarities of crocodilian population dynamics and their possible importance for management strategies. In *Crocodiles: Proceedings of the 7th Working Meeting of the Crocodile Specialist Group,* 434–442. IUCN Publications N.S., Caracas, Venezuela.

Magnusson, W. E., and A. P. Lima. 1991. The ecology of a cryptic predator, *Paleosuchus trigonatus,* in a tropical rainforest. *Journal of Herpetology* 25: 41–48.

Magnusson, W. E., A. P. Lima, and R. A. Sampaio. 1985. Sources of heat for nests from *Paleosuchus trigonatus* and a review of crocodilian nest temperatures. *Journal of Herpetology* 19: 199–207.

Malone, B. 1979. The systematics, phylogeny, and paleobiology of the genus Alligator. PhD diss., City University of New York, New York. 144 pp.

Marks, R. B. 1998. *Tigers, Rice, Silk, and Silt.* Cambridge University Press, New York. 383 pp.

Markwick, P. J. 1998. Fossil crocodilians as indicators of Late Cretaceous and Cenozoic climates: Implications for using paleontological data in reconstructing paleoclimate. *Palaeogeography, Palaeoclimate, Palaeoecology* 137: 205–271.

Mathews, F., D. Moro, R. Strachan, M. Gelling, and N. Buller. 2006. Health surveillance in wildlife reintroductions. *Biological Conservation* 131: 338–347.

McAliley, L. R., R. E. Willis, D. A. Ray, P. S. White, C. A. Brochu, and L. D. Densmore III. 2006. Are crocodiles really monophyletic? Evidence for subdivisions from sequence and morphological data. *Molecular Phylogenetics and Evolution* 39: 16–32.

McGinnity, P., P. Prodöhl, A. Ferguson, R. Hynes, N. O. Maoiléidigh, N. Baker, D. Cotter, B. O'Hea, D. Cooke, G. Rogan, J. Taggart, and T. Cross. 2003. Fitness reduction and potential extinction of wild populations of Atlantic salmon, *Salmo salar,* as a result of interactions with escaped farm salmon. *Proceedings of the Royal Society of London, Series B, Biological Sciences* 270: 244–250.

McIlhenny, E. A. 1935. *The Alligator's Life History.* Christopher Publishing House, Boston. 177 pp.

McNease, L., and T. Joanen. 1979. Distribution and relative abundance of the alligator in Louisiana coastal marshes. *Proceedings of the Annual Conference of Southeast Association of Game and Fish Commissions* 32: 182–186.

Mead, J. I., D. W. Steadman, S. H. Bedford, C. J. Bell, and M. Spriggsa. 2002. New extinct Mekosuchine crocodile from Vanuatu, South Pacific. *Copeia* 2002: 632–641.

Medem, F. 1981. *Los Crocodylia de Sur America.* Vol. 1, *Los Crocodylia de Colombia.* COLCIENCIAS, Bogota, Colombia. 354 pp.

————. 1983. *Los Crocodylia de Sur America.* Vol. 2. COLCIENCIAS, Bogota, Colombia. 270 pp.

Meyer, G. 1976. Alligator ecology and population structure on Georgia Sea Islands. Master's thesis, New York University, New York, N.Y.

Miller, B. J., R. P. Reading, and S. C. Forrest. 1996. *Prairie Night: Black-footed Ferrets and the Recovery of Endangered Species.* Smithsonian Institution Press, Washington, D.C. 254 pp.

Minton, S. A., and M. R. Minton. 1973. *Giant Reptiles.* Scribner, New York. 345 pp.

Mo, X. Q., T. J. Zhao, and P. C. Qin. 1991. The origin of the Chinese alligator. *Science in China, Series B* 10: 1047–1053. (In Chinese.)

Molnar, R. E., T. Worthy, and P. M. A. Willis. 2002. An extinct Pleistocene endemic Mekosuchine crocodylian from Fiji. *Journal of Vertebrate Paleontology* 22: 612–628.

Mundkur, B. 1978. The roots of ophidian symbolism. *Ethos* 6: 125–158.

Murray, G., and I. G. Cook. 2003. *Green China Seeking Ecological Alternatives*. Routledge Curzon, New York. 272 pp.

National Environmental Protection Agency of China. 1997. *China's National Report on Implementation of the Convention on Biological Diversity*. National Environmental Protection Agency of China, Beijing. 152 pp.

Neill, W. T. 1971. *The Last of the Ruling Reptiles*. Columbia University Press, New York. 486 pp.

Njau, J. K., and R. J. Blumenschine. 2006. A diagnosis of crocodile feeding traces on larger mammal bone, with fossil examples from the Plio-Pleistocene Olduvai Basin, Tanzania. *Journal of Human Evolution* 50: 142–162.

Öhman, A., A. Flyky, and F. Esteves. 2001. Emotion drives attention: Detecting the snake in the grass. *Journal of Experimental Psychology: General* 130: 466–478.

Osborne, A. 1998. Highlands and lowlands: Economic and ecological interactions in the lower Yangzi region under the Qing. In *Sediments of Time*, ed. M. Elvin and T. Liu, 203–234. Cambridge University Press, New York.

Palmer, M. L., and F. J. Mazzotti. 2004. Structure of Everglades alligator holes. *Wetlands* 24: 115–122.

Pearson, R., and A. Underhill. 1987. The Chinese Neolithic: Recent trends in research. *American Anthropologist* 89: 807–822.

Pellegrin, J. 1937. Mort d'un alligator presume avoir vecu 85 ans a la menagerie des reptiles. *Bulletin, Museum National d'Histoire Natural* 2nd Ser. 9: 176–177.

Perdue, P. C. 1987. *Exhausting the Earth: State and Peasant in Hunan, 1500–1850*. Harvard University Press, Cambridge, Mass. 400 pp.

Perkins, D. H. 1969. *Agricultural Development in China, 1368–1968*. Aldine, Chicago. 395 pp.

Phelps, M. 1804. *Memoirs and Adventures of Captain Matthew Phelps*. Anthony Haswell, Bennington, Vt. 210 pp.

Piña, C., A. Larriera, and M. R. Cabrera. 2003. Effect of incubation temperature on incubation period, sex ratio, hatching success, and survivorship in *Caiman latirostris* (Crocodylia, Alligatoridae). *Journal of Herpetology* 37: 199–202.

Platt, S. G. 2000. Dens and denning behavior of Morelet's crocodile (*Crocodylus moreletii*). *Amphibia-Reptilia* 21: 232–237.

Pooley, A. C. 1969. The burrowing behaviour of crocodiles. *Lammergeyer* 10: 60–63.

Pope, C. H. 1935. *The Reptiles of China*. Vol. 10, *Natural History of Central Asia*. C. A. Reeds, New York. 604 pp.

———. 1940. *China's Animal Frontier*. Viking Press, New York. 192 pp.

———. 1955. *The Reptile World*. Alfred A. Knopf, New York. 325 pp.

Porter, De. L. 1996. *From Deluge to Discourse: Myth, History, and the Generation of Chinese Fiction*. State University of New York Press, Albany. 284 pp.

Pough, F. H. 1980. The advantages of ectothermy for tetrapods. *American Naturalist* 115: 92–112.

PRC. 1992. Registration of the Anhui Research Centre of Chinese Alligator Reproduction for *Alligator sinensis*. Proposal to CITES. Kyoto, Japan.

Qian, H. 2007. Relationships between plant and animal species richness at a regional scale in China. *Conservation Biology* 21: 937–944.

Quammen, D. 2003. *Monster of God: The Man-Eating Predator in the Jungles of History and the Mind*. Norton, New York. 515 pp.

Raffaele, P. 2008. Forbidden no more. *Smithsonian* 38: 80–90.

Reed, D., H., J. J. O'Grady, B. W. Brook, J. D. Ballou, and R. Frankham. 2003. Estimates of minimum viable population sizes for vertebrates and factors influencing those estimates. *Biological Conservation* 113: 23–34.

Reese, A. M. 1915. *The Alligator and Its Allies*. Putnam's, New York. 358 pp.

Reeves, R. R., B. D. Smith, E. A. Crespo, and G. Notarbartolo di Sciara (compilers). 2003. *Dolphins, Whales, and Porpoises: 2002–2010 Conservation Action Plan for the World's Cetaceans*. IUCN/SSC Cetacean Specialist Group. IUCN–World Conservation Union, Gland, Switzerland, and Cambridge, Eng. http://data.iucn.org/dbtw-wpd/edocs/2003-009.pdf.

Roos, J., R. K. Aggarwal, and A. Janke. 2007. Extended mitogenomic phylogenetic analyses yield new insight into crocodylian evolution and their survival of the Cretaceous-Tertiary boundary. *Molecular Phylogenetics and Evolution* 45: 663–673.

Rootes, W. L., and R. H. Chabreck. 1993. Reproductive status and movement of adult female alligators. *Journal of Herpetology* 27: 121–126.

Rootes, W. L., R. H. Chabreck, V. L. Wright, B. W. Brown, and T. J. Hess. 1991. Growth rates of American alligators in estuarine and palustrine wetlands in Louisiana. *Estuaries* 14: 489–494.

Ross, C. A., and C. H. Ernst. 1994. *Alligator mississippiensis* (Daudin) American Alligator. *Catalogue of American Amphibians and Reptiles* 600.1-600.14.

Ross, J. P. (ed.). 1998. *Crocodiles: Status Survey and Conservation Action Plan.* 2nd Edition. IUCN/SSC Crocodile Specialist Group. IUCN–World Conservation Union, Gland, Switzerland, and Cambridge, Eng. 96 pp.

Ross, J. P., M. Cherkiss, and F. Mazzotti. 2000. Problems of success: Conservation consequences of crocodilian-human conflict. In *Crocodiles. Proceedings of the 15th Working Meeting of the Crocodile Specialist Group*, 442–445. IUCN– World Conservation Union. Gland, Switzerland.

Sanderson, E. W., K. H. Redford, C-L. B. Chetkiewicz, R. A. Medellin, A. R. Rabinowitz, J. G. Robinson, and A. B. Taber. 2002. Planning to save a species: The jaguar as a model. *Conservation Biology* 16: 58–72.

Schaller, G. 1993. *The Last Panda.* University of Chicago Press, Chicago. 356 pp.

Schapiro, J. 2001. *Mao's War against Nature: Politics and the Environment in Revolutionary China.* Cambridge University Press, New York. 334 pp.

Schmidt, K. P. 1927. Notes on Chinese reptiles. *Bulletin of the American Museum of Natural History.* 54: 467–551.

Scott, D. A. (compiler). 1989. *A Directory of Asian Wetlands.* IUCN–World Conservation Union, Gland, Switzerland. 1181 pp.

Seddon, P. J., D. P. Armstrong, and R. F. Maloney. 2007. Developing the science of reintroduction biology. *Conservation Biology* 21: 303–312.

Seebacher, F., R. M. Elsey, and P. L. I. Trosclair. 2003. Body temperature null distributions in reptiles with nonzero heat capacity: Seasonal thermoregulation in the American alligator (*Alligator mississippiensis*). *Physiological and Biochemical Zoology* 76: 348–359.

Shaffer, M. L. 1987. Minimum viable populations: Coping with uncertainty. In *Viable Populations for Conservation*, ed. M. E. Soulé, 69–86. Cambridge University Press, Cambridge, Eng.

Shaffer, M. L., L. H. Watchman, W. J. Snape III, and I. K. Latchis. 2002. Population viability analysis and conservation policy. In *Population Viability Analysis*, ed. S. R. Beissinger and D. R. McCullough, 123–142. University of Chicago Press, Chicago.

Shankman, D., and Q. Liang. 2003. Landscape changes and increasing flood frequency in China's Poyang Lake region. *Professional Geographer* 55: 434–445.

Shao, M., X. Tang, Y. Zhang, and W. Li. 2006. City clusters in China: Air and surface water pollution. *Frontiers in Ecology and the Environment* 4: 353–361

Shao, W. 2000. The Longshan period and incipient Chinese civilization. *Journal of East Asian Archaeology* 2: 195–226.

Shelach, G. 2000. The earliest Neolithic cultures of northeast China: Recent discoveries and new perspectives on the beginning of agriculture. *Journal of World Prehistory* 14: 363–413.

Shi, K., X. Zhou, and R. Wang. 2006. *The Story of the Chinese Alligator.* Chinese Forestry Publishing House, Beijing. 72 pp.

Shi, Y., Z. Kong, S. Wang, L. Tang, F. Wang, T. Yao, X. Zhao, P. Zhang, and S. Shi. 1993. Mid-Holocene climates and environments in China. *Global Planetary Change* 7: 219–233.

Shine, R., 1988. Parental care in reptiles, In *Biology of the Reptilia.* Vol. 16B, *Defense and Life History*, ed. C. Gans and R. B. Huey, 275–329. Alan R. Liss, New York.

———. 1999. Why is sex determined by nest temperature in many reptiles? *Trends in Ecology and Evolution* 14: 186–189.

Shine, T., W. Bohme, H. Nickel, D. F. Thies, and T. Wilms. 2001. Rediscovery of relict populations of the Nile crocodile *Crocodylus niloticus* in southeastern Mauritania, with observations on their natural history. *Oryx* 35: 260–262.

Smith, A. T., and Y. Xie (eds.) 2008. *A Guide to the Mammals of China.* Princeton University Press, Princeton, N.J. 544 pp.

Smith, G. E. 1922. *The Evolution of the Dragon.* Manchester University Press, Manchester, Eng. 234 pp.

Smith, H. M. 1893. Notes on the alligator industry. *Bulletin of the U.S. Fisheries Commission* 11: 343–345.

Smith, N. J. H. 1981. Caimans, capybaras, otters, manatees, and man in Amazonia. *Biological Conservation* 19: 177–187.

Snyder, D. 2001. Evolution of the modern *Alligator* lineage. Master's thesis, University of Florida, Gainesville. 102 pp.

———. 2007. Morphology and systematics of two Miocene alligators from Florida, with a discussion of Alligator biogeography. *Journal of Paleontology* 81: 917–928.

Snyder, N. F. R., S. R. Derrickson, S. R. Beissinger, J. W. Wiley, T. B. Smith, W. D. Toone, and B. Miller. 1996. Limitations of captive breeding in endangered species recovery. *Conservation Biology* 10: 338–348.

Soorae, P. S., and M. R. Stanley Price. 1998. General introduction to re-introductions. In *Crocodiles: Proceedings of the 14th Working Meeting of the Crocodile Specialist Group,* 268–284. IUCN–World Conservation Union, Gland, Switzerland, and Cambridge, Eng.

Soulé, M. E. 1980. Thresholds for survival: Maintaining fitness and evolutionary potential. In *Conservation Biology: An Evolutionary-Ecological Perspective,* ed. M. E. Soulé and B. A. Wilcox, 151–170. Sinauer, Sunderland, Mass.

———. 1987. *Viable Populations for Conservation.* Island Press, Washington, D.C. 189 pp.

Sowerby, A. de C. 1936. Pangolin and alligators in Shanghai. *China Journal* 25: 110–111.

———. 1943. Amphibians and reptiles recorded from or known to occur in the Shanghai area. *Notes D'Herpétologie, Musée Heude, Université L'Aurore, Shanghai* 1: 1–16.

Stanley Price, M. R., and P. S. Soorae. 2003. Reintroductions: Whence and whither? *International Zoo Yearbook* 38: 61–75

State Environmental Protection Administration. 2005. *China's Third National Report on Implementation of the Convention on Biological Diversity.* China Environmental Science Press, Beijing. 110 pp.

State Forestry Administration. 1998. State Forestry Administration, *Department of Forestry Newsletter,* June 1998, 6–7. (In Chinese.)

Steel, R. 1973. *Handbuch der Paleoherpetologie,* Pt. 16, *Crocodylia.* Gustav Fischer Verlag, Portland, Ore. 116 pp.

Sterckx, R. 2000. Transforming the beasts: Animals and music in early China. *T'oung Pao* 86: 1–46.

———. 2005. Animal classification in ancient China. *East Asian Science, Technology, and Medicine* 23: 26–53.

Stirrat, S., D. Lawson, W. J. Freeland, and R. Morton. 2001. Monitoring *Crocodylus porosus* in the Northern Territory of Australia: A retrospective power analysis. *Wildlife Research* 28: 547–554.

Swinhoe, R. 1870. Notes on reptiles and batrachians collected in various parts of China. *Proceeding of the Zoological Society of London* 1870: 409–412.

Tao, J., M. Chen, and S. Xu. 2006. A Holocene environmental record from the southern Yangtze River delta, Eastern China. *Palaeogeography, Palaeoclimate, Palaeoecology* 230: 204–229.

Tchernov, E. 1986. *Evolution of the Crocodiles in East and North Africa.* Centre National de la Recherché Scientifique, Paris. 65 pp.

Thorbjarnarson, J. B. 1991. An analysis of the spectacled caiman (*Caiman crocodilus*) harvest program in Venezuela. In *Neotropical Wildlfe Use and Conservation,* ed. J. G. Robinson and K. H. Redford, 217–235. University of Chicago Press, Chicago.

———. 1993. Diet of the spectacled caiman (*Caiman crocodilus*) in the central Venezuelan llanos. *Herpetologica* 49: 108–117.

———. 1994. Reproductive ecology of the spectacled caiman (*Caiman crocodilus*) in the Venezuelan llanos. *Copeia* 1994: 907–918.

———. 1996. Reproductive characteristics of the Crocodylia. *Herpetologica* 52: 8–24.

Thorbjarnarson, J. B., and R. Da Silveira. 2000. Secrets of the flooded forest. *Natural History* 109: 70–79.

Thorbjarnarson, J., and G. Hernandez. 1993. Reproductive ecology of the Orinoco crocodile (*Crocodylus intermedius*) in Venezuela. 2. Reproductive and social behavior. *Journal of Herpetology* 27: 371–379.

Thorbjarnarson, J., and A. Velasco. 1999. Economic incentives for management of Venezuelan caiman. *Conservation Biology* 13: 397–406.

Thorbjarnarson, J., X. Wang, S. Ming, L. He, Y. Ding, and S. T. McMurry. 2001. Wild populations of the Chinese alligator approach extinction. *Biological Conservation* 103: 93–102.

Tilson, R., H. Defu, J. Muntifering, and P. J. Nyhus. 2004. Dramatic decline of wild South China tigers *Panthera tigris amoyensis*: Field survey of priority tiger reserves. *Oryx* 38: 40–47.

Tong, E. 2003. Magicians, magic, and shamanism in ancient China. *Journal of East Asian Archaeology* 4: 27–73.

Topsell, E. 1658. *The Historie of Four-footed Beasts and Serpents*. Vol. 2, *The Historie of Serpents*. E. Cotes for G. Sawbridge, London. 246 pp.

Trauth, S. E., and M. L. McCallum. 2001. *Alligator mississippiensis* (American alligator). Winter mortality. *Herpetological Review* 32: 250–251.

Tucker, A. D. 1997. Validation of skeletochronology to determine age of freshwater crocodiles (*Crocodylus johnstoni*). *Marine and Freshwater Research* 48: 343–351.

Tucker, A. D., C. J. Limpus, K. R. McDonald, and H. McCallum. 2006. Growth dynamics of freshwater crocodiles (*Crocodylus johnstoni*) in the Lynd River, Queensland. *Australian Journal of Zoology* 54: 409–415.

Turvey, S. T., R. L. Pitman, B. L. Taylor, J. Barlow, T. Akamatsu, L. A. Barrett, X. Zhao, R. R. Reeves, B. S. Stewart, K. Wang, Z. Wei, X. Zhang, L. T. Pusser, M. Richlen, J. R. Brandon, and D. Wang. 2007. First human-caused extinction of a cetacean species? *Biology Letters* 3: 537–540.

Underhill, A. P. 1997. Current issues in Chinese Neolithic archaeology. *Journal of World Prehistory* 11: 103–160.

Van Slyke, L., 1988. *Yangtze: Nature, History, and the River*. Addison-Wesley, New York. 211 pp.

Vermeer, E. B. 1998. Population and ecology along the frontier in Qing China. In *Sediments of Time*, ed. M. Elvin and T. Liu, 235–279. Cambridge University Press, New York.

Vliet, K. 1989. Social displays of the American alligator (*Alligator mississippiensis*). *American Zoologist* 29: 1019–1031.

Vörösmarty, C. J., B. Fekete, and B. A. Tucker. 1996. River Discharge Database, Version 1.0 (RivDIS v1.0), Volumes 0 through 6. A contribution to IHP-V Theme 1. Technical Documents in Hydrology Series. UNESCO, Paris.

Wan, H. 2003. Policies and measures on flood mitigation in China since 1998. International Conference on Total Disaster Risk Management, 2–4 December 2003, Kobe, Japan. Kobe: Asian Disaster Reduction Center. http://web.adrc.or.jp/publications/TDRM2003Dec/11_MR.%20HONGTAO%20WAN%20_FINAL_.pdf.

Wan, Z. 1997. Crocodile conservation, captive breeding, and utilization in the world. *Chinese Wildlife* 18: 35–37.

Wan, Z., C. Gu, X. Wang, and C. Wang. 1998. Conservation, management, and farming of crocodiles in China. In *Crocodiles: Proceedings of the 14th Working Meeting of the Crocodile Specialist Group*, 80–100. IUCN–World Conservation Union, Gland, Switzerland, and Cambridge, Eng.

Wang, D. 1988. *Long Feng Wenhua Yuanliu* (Origins of the Dragon and Phoenix Culture). Gongyi Meishu Chubanshe, Beijing.

Wang, G., L. He, and M. Shao. 2000. Relationship between egg hatching and environment in wild *Alligator sinensis* in Jinxiang county. *Sichuan Journal of Zoology* 19: 82–83. (In Chinese.)

Wang, H., Z. Wang, Z. Wang, J. Gong, and X. Wang. 2006. Study on behaviorally thermo-regulation of Chinese alligator under artificial feeding condition. *Chinese Journal of Zoology* 6: 61–66. (In Chinese with English abstract.)

Wang, J., and Z. Huang, 1996. Nile crocs in China. *Crocodile Specialist Group Newsletter* 15: 6.

———. 1997. Chiangxiang conservation area for Chinese alligator. *Crocodile Specialist Group Newsletter* 16: 11–12.

Wang, Q. 2006. A study of the sustainable development on ecotourism in Xiazhu Lake of Deqing County. *Newsletter of Zhejiang.* 3: 27–31. (In Chinese.)

Wang, R., and X. Ma. 2000. The formation of Chinese alligator egg band and its practical value. *Chinese Journal of Zoology* 35: 22–25. (In Chinese.)

Wang, R., R. Ye, Y. Zhou, X. Wu, and T. Xia. 2006. Growth regularity in captive Chinese alligator, *Alligator sinensis*. *Acta Hydrobiologica Sinica* 30: 593–600. (In Chinese with English abstract.)

Wang, W. 2001. Protection and management project of the existing wild population of *Alligator sinensis* in Anhui Province. In *Status Quo and Future of Conservation for Chinese Alligator and Crocodiles in the World: Proceedings of the International Workshop on Conservation and Reintroduction of Chinese Alligator, Hefei, China, 2001, and International Workshop on Captive Breeding and Commerce Management in Crocodylia, Guangzhou, China, 2001,* ed. W. Wang, X. Ruan, P. Si, D. Zhang, Y. Luo, C. Wang, Q. Sun, F. Qian, H. Jiang, 41–47. China Forestry Publishing House. Beijing.

Wang, X., D. Wang, X. Wu, R. Wang, and C. Wang. 2006. Congregative effect of Chinese alligator's bellowing chorus in mating season and its function in reproduction. *Acta Zoologica Sinica* 52: 663–668. (In Chinese with English abstract.)

———. 2007. Acoustic signals of Chinese alligators (*Alligator sinensis*): Social communication. *Journal of the Acoustical Society of America* 121: 2984–2989.

Wang, Y. 1962. The Yangtze alligator. *China Reconstructs* March 1962: 38–39.

Wang, Y., W. Zhu, and C. Wang. 2003. D-loop sequence variation of mitochondrial DNA in captive Chinese alligator. *Acta Genetica Sinica* 30: 425–430.

Wang, Y., W. Zhu, L. Huang, K. Zhou, and R. Wang. 2006. Genetic diversity of Chinese alligator (*Alligator sinensis*) revealed by AFLP analysis: An implication on the management of captive conservation. *Biodiversity and Conservation* 15: 2945–2955.

Watanabe, M. E. 1981. The Chinese alligator, *Alligator sinensis* Fauvel, in the People's Republic of China: Distribution, status, conservation, and mating behavior and suggestions for future conservation. Typescript. U.S. Fish and Wildlife Service, Order no. 98210-1120. 33 pp.

———. 1982. The Chinese alligator: Is farming the last hope? *Oryx* 174: 176–181.

Watters, L., and X. Wang. 2002. The protection of wildlife and endangered species in China. *Georgetown International Environmental Law Review* 14: 489–525.

Webb, G. J. W. 2000. Sustainable use of large reptiles: An introduction to issues. In *Crocodiles: Proceedings of the 15th Working Meeting of the Crocodile Specialist Group,* 413–430. IUCN–World Conservation Union, Gland, Switzerland, and Cambridge, Eng.

Webb, G. J. W., A. M. Beal, S. C. Manolis, and K. E. Dempsey. 1987. The effects of incubation temperature on sex determination and embryonic development rate in *Crocodylus johnstoni* and *C. porosus*. In *Wildlife Management: Crocodiles and Alligators,* ed. G. J. W. Webb, S. C. Manolis and P. Whitehead, 507–531. Surrey Beatty and Sons, Chipping Norton, Australia.

Webb, G. J. W., A. R. C. Britton, S. C. Manolis, B. Ottley, and S. Stirrat. 2000. The recovery of *Crocodylus porosus* in the Northern Territory of Australia: 1971–1998. In *Crocodiles: Proceedings of the 15th Working Meeting of the Crocodile Specialist Group,* 195–234. IUCN–World Conservation Union, Gland, Switzerland, and Cambridge, Eng.

Webb, G. J. W., R. Buckworth, and S. C. Manolis. 1983. *Crocodylus johnstoni* in the McKinlay River area, N.T. 6: Nesting biology. *Australian Wildlife Research* 10: 607–637.

Webb, G. J. W., and H. Cooper-Preston. 1989. Effects of incubation temperature on crocodiles and the evolution of reptilian oviparity. *AmericanZoologist* 29: 953–972.

Webb, G. J. W., S. C. Manolis, and R. Buckworth. 1983. *Crocodylus johnstoni* in the McKinlay river area, N.T. 1: Variation in the diet, and a new method of assessing the relative importance of prey. *Australian Journal of Zoology* 30: 877–899.

Webb, G. J. W., and B. Vernon. 1992. Crocodilian management in the People's Republic of China: A review with recommendations. In *Crocodile Conservation Action*, 1–27. Special Publication of the Crocodile Specialist Group of the Species Survival Commission of the IUCN–World Conservation Union, Gland, Switzerland.

Weigelt, J. 1989. *Recent Vertebrate Carcasses and Their Paleobiological Implications*. University of Chicago Press, Chicago. 188 pp.

Weldon, P. J., and M. W. J. Ferguson. 1993. Chemoreception in crocodilians: Anatomy, natural history, and empirical results. *Brain Behavior and Evolution* 41: 239–245.

Wen, H. 1980. The preliminary analysis of the distribution and the cause of change of *Crocodylus porosus* of China in the historical periods. *Journal of Shanghai Normal University (Natural Sciences)* 1980(3): 112–122.

———. 1981. Analysis of historical geographic changes of Chinese alligator. *Journal of Xiangtan University* (Natural Sciences). 1981: 112–122. (In Chinese.)

———. 1995. Historical changes in the distribution of the Chinese alligator. In *The Shift of Plants and Animals Distribution in the Historical Period*, ed. R. Wen, 160–162. Chongqing Publishing House, Chongqing. (In Chinese.)

Wen, J. 1999. Evolution of eastern Asian and eastern North American disjunct distributions in flowering plants. *Annual Review of Ecology and Systematics* 30: 421–455.

Werner, E. T. C. 1928. *Autumn Leaves: A Sheaf of Papers, Sociological and Sinological, Philosophical and Metaphysical with an Autobiographic Note*. Kelly and Walsh, Shanghai. 747 pp.

White, J. 2000. Bites and stings from venomous animals: A global overview. *Therapeutic Drug Monitoring* 22: 65–68.

White, T. H., Jr., J. A. Collazo, and F. J. Vilella. 2005. Survival of captive-reared Puerto Rican Parrots released in the Caribbean National Forest. *Condor* 107: 424–432.

Wilkinson, P. M. 1984. *Nesting Ecology of the American Alligator in Coastal South Carolina*. South Carolina Wildlife and Marine Research Department, Division of Wildlife and Freshwater Fisheries, Charleston, S.C. 113 pp.

Wilkinson, P. M., and W. E. Rhodes. 1997. Growth rates of American alligators in coastal South Carolina. *Journal of Wildlife Management* 61: 397–402.

Wilson, E. O. 1996. *In Search of Nature*. Island Press. Washington, D.C. 229 pp.

Wink, C. S., and R. M. Elsey. 1986. Changes in femoral morphology during egg-laying in *Alligator mississippiensis*. *Journal of Morphology* 189: 183–188.

Winkler, M. G., and P. K. Wang. 1993. The Late-quaternary vegetation and climate of China. In *Global Climates since the Last Glacial Maximum*, ed. H. E. Wright Jr., J. E. Kutzbach, T. Webb III, W. F. Ruddiman, F. A. Street-Perrott, and P. J. Bartline, 221–264. University of Minnesota Press, Minneapolis.

Woodward, A. R., and C. T. Moore. 1992. *Alligator Age Determination*. Final Report Study Number 763. Bureau of Wildlife Research, Division of Wildlife, Florida Game and Fresh Water Fish Commission, Gainesville, Fla.

Woodward, A. R., C. T. Moore, and M. F. Delany. 1992. *Experimental Alligator Harvest*. Final Report. Study Number 7567. Florida Game and Freshwater Fish Commission, Gainesville, Fla.

Woodward, A. R., J. H. White, and S. B. Linda. 1995. Maximum size of the alligator (*Alligator mississippiensis*). *Journal of Herpetology* 29: 507–513.

Wu, J., and X. Wang. 2004. Regulation of bellowing of Chinese alligators (*Alligator sinensis*) in the wild. *Zoological Research* 25: 281–286. (In Chinese with English abstract.)

Wu, X., Y. Wang, K. Zhou, W. Zhu, J. Nie, and C. Wang. 2003. Complete mitochondrial DNA sequence of Chinese alligator, *Alligator sinensis*, and phylogeny of crocodiles. *Chinese Science Bulletin* 48: 2050–2054.

Wu, X., Y. Wang, K. Zhou, W. Zhu, J. Nie, C. Wang, and W. Xie. 2002. Genetic variation in captive population of Chinese alligator, *Alligator sinensis*, revealed by random amplified polymorphic DNA (RAPD). *Biological-Conservation* 106: 435–441.

Xia, T., and X. Wu. 2005. Burrowing characteristic of Chinese alligators under high density in captivity. *Acta Herpetologica Sinica* 10: 164–170 (In Chinese with English abstract.)

Xinhua News Agency. 2004. China looks to take hybrid rice global for world food security. http://en.ce.cn/National/Science/t20040415_664253.shtml.

Xu, H., and R. H. Giles. 1995. A view of wildlife management in China. *Wildlife Society Bulletin* 23: 18–25.

Xu, J., M. Bennett, R. Tao, and J. Xu. 2004. China's Sloping Land Conversion Program four years on: Current situation, pending issues. *International Forestry Review* 6: 317–326.

Xu, J., and D. R. Melick. 2007. Rethinking the effectiveness of public protected areas in southwestern China. *Conservation Biology* 21: 318–328.

Xu, J., W. Xie, and H. Pan. 1990. Nesting ecology of Chinese alligator in artificial reproductive site. *Sichuan Journal of Zoology* 8: 30–31.

Xu, Q., and C. Huang. 1984. Some problems in evolution and distribution of *Alligator*. *Vetebrata PalAsiatica* 22: 49–53. (In Chinese with English abstract.)

Yang, J., Y. Wang, V., R. A. Spicer, Mosbrugger, C. Li, and Q. Sun. 2007. Climatic reconstruction at the Miocene Shanwang Basin, China, using leaf margin analysis, CLAMP, co-existence approach, and overlapping distribution analysis. *American Journal of Botany* 94: 599–608.

Yang, X., and J. Pang. 2006. Implementing China's "Water Agenda 21." *Frontiers in Ecology and the Environment* 4: 362–368.

Yardley, J. 2004. In a tidal wave, China's masses pour from farm to city. *New York Times*, 12 September 2004. ("Week in Review" p. 6.)

Yin, H., G. Liu, J. Pi, G. Chen, and C. Li. 2007. On the river-lake relationship of the middle Yangtze reaches. *Geomorphology* 85: 197–207.

Yoshinobu, S. 1998. Environment versus water control: The case of the southern Hangzhou Bay area from the Mid-Tang through the Qing. In *Sediments of Time*, ed. M. Elvin and T. Liu, 135– 164. Cambridge University Press, New York.

Zhang, F., Y. Li, Z. Guo, and B. R. Murray. 2009. Climate warming and reproduction in Chinese alligators. *Animal Conservation* 12: 128–137.

Zhang, F., X. Wu, and J. Zhu. 2006. Ecology on the making nest and laying eggs of Chinese alligator (*Alligator sinensis*) under artificial feeding conditions. *Zoological Research* (China) 27: 151–156.

Zhang, J., and X. Wang. 1998. Notes on recent discovery of ancient cultivated rice at Jiahu, Henan Province: A new theory concerning the origin of *Oryza japonica* in China. *Antiquity* 72: 897–901.

Zhang, M., and Z. Huang. 1979. Chinese crocodilian species: The Chinese alligator and the saltwater crocodile. *Dongu Zazhi* 2: 52.

Zhang, X., D. R. Wang, Z. Liu, Y. Wei, Y. Hua, Z. Wang, Z. Chen, and L. Wang. 2003. The Yangtze River dolphin or baiji (*Lipotes vexillifer*): Population status and conservation issues in the Yangtze River, China. *Aquatic Conservation of Marine and Freshwater Ecosystems* 13: 51–64

Zhang, Z. D. 1989. A major research achievement in captive reproduction of Chinese alligators. *Chinese Herpetological Research* 2; 69–71.

———. 1992. Developments in alligator farming in China. *Crocodile Specialist Group Newsletter* 11: 5.

———. 1995. Captive breeding Chinese alligator. *Crocodile Specialist Group Newsletter* 14: 9

———. 1996. Tourist development for Chinese alligators. *Crocodile Specialist Group Newsletter* 15: 5–6.

Zhang, Z., J. Diren, Y. Zhao, and H. Pan. 1985. The growth rates of young Chinese alligators in captivity. *Acta Herpetologica Sinica* 5: 217–222. (In Chinese with English abstract.)

Zhang, Z., and B. Wang. 1987. The growth of young Chinese alligator during hibernation. *Acta Herpetologica Sinica* 6: 16–20. (In Chinese with English abstract.)

Zhao, E., and K. Adler. 1993. *Herpetology of China*. Contributions to Herpetology, no. 10. Society for the Study of Amphibians and Reptiles, Oxford, Ohio. 522 pp.

Zhao, K., Y. Zong, and J. Ma. 1986. On the ancient crocodiles of Guangdong Province. *Acta Herpetologica Sinica* 5: 161–165.

Zhao, Q. 1992. *A Study of Dragons, East and West*. Peter Lang, New York. 237 pp.

Zhao, S., J. Fang, S. Miao, B. Gu, S. Tao, C. Peng, and Z. Tang. 2005. The 7-decade degradation of a large freshwater lake in Central Yangtze River, China. *Environmental Science and Technology* 39: 431–436.

Zhao, Z. 1998. The middle Yangtze region in China is one place where rice was domesticated: Phytolith evidence from the Diaotonghuan Cave, northern Jiangxi. *Antiquity* 72: 885–897.

Zhou, B. 1982. Skeletal remains of the Yangtze alligator discovered at the Wangyin Neolithic site, Yanzhou, Shandong. *Archaoelogica Sinica* 2: 251–260. (In Chinese.)

Zhou, Y. J. 1997. Analysis of the decline of the wild *Alligator sinensis* population. *Sichuan Journal of Zoology* 16: 137. (In Chinese.)

Zhu, G. 2002. Historical development and improvement of China's nature reserves. In *Proceedings of IUCN/WCPA-EA-4 Taipei Conference*, 51–55. IUCN–World Conservation Union, Taipei, Taiwan.

Zhu, H. 1997. Observations on a wild population of *Alligator sinensis*. *Sichuan Journal of Zoology* 16: 40–41. (In Chinese.)

Zong, Y., and X. Chen. 2000. The 1998 flood on the Yangtze, China. *Natural Hazards* 22: 165–184.

Index